COMPLETE
Jams, Preserves
AND Chutneys

COMPLETE
Jams, Preserves AND Chutneys

Sara Paston-Williams

First published in Great Britain in 1985 by David & Charles under the title *The Country House Kitchen Store Cupboard*. Revised edition published in Great Britain in 1999 by The National Trust Limited under the title *Jams, Preserves and Edible Gifts* and in 2008 under the title *Good Old-Fashioned Jams, Preserves and Chutneys*.

This edition published in the United Kingdom in 2015 by
National Trust Books
1 Gower Street
London WC1E 6HD

An imprint of Pavilion Books Ltd

Text and recipes © Sara Paston-Williams
Volume © National Trust Books

The moral right of the author has been asserted.

All rights reserved. No part of this publication may be reproduced, stored in a retrieval system, or transmitted in any form or by any means, electronic, mechanical, photocopying, recording or otherwise, without the prior written permission of the copyright owner.

ISBN 9781909881518

A CIP catalogue record for this book is available from the British Library.

10 9 8 7 6 5 4 3 2 1

Reproduction by Mission Productions, Hong Kong
Printed and bound by Toppan Leefung, China

Design by Rosamund Saunders
Home economy by Jane Suthering
Photography by Tara Fisher, other than © National Trust Images/Mark Bolton: 101; NTI/James Dobson: 141; NTI/Myles New: 171; NTI/Stephen Robson: 128; NTI/Arnhel de Sera: 155; NTI/Ian West: 103 and ©StockFood: 27, 33, 46, 51, 80, 85, 93, 97, 137, 187.

This book can be ordered direct from the publisher at the website: www.pavilionbooks.com, or try your local bookshop. Also available at National Trust shops.

The National Trust for Places of Historic Interest or Natural Beauty, National Trust (Enterprises) Ltd and Pavilion Books are not responsible for the outcome of any recipe or other process in this recipe book and other than for death or personal injury caused by our negligence we do not accept any liability for any loss or damage howsoever caused.

Warning: recipes containing raw eggs are unsuitable for pregnant women or young children.

While we use all reasonable endeavours to ensure the accuracy of each recipe, you may not always achieve the desired results due to individual variables such as ingredients, cooking temperatures and equipment used, or cooking abilities.

Contents

Introduction 6

Chutneys, Relishes and Sauces 8

Pickles 54

Flavoured Sugars, Salts and Syrups 86

Jams and Jellies 106

Fruit Butters, Cheeses, Pastilles, Leathers and Curds 148

Marmalades 170

Mincemeats 184

Fruits Preserved in Alcohol 192

Fruit and Flower Drinks 204

Index 224

Introduction

There are few activities more enjoyable than making your own preserves. As modern life moves faster and faster, it is easy to lose sight of the good things in life: the perfume of bubbling fruit, the glistening of a pot of hedgerow jelly and the crisp bite of a homemade pickled onion. Whether you are an experienced cook or rarely approach the stove, making preserves – especially chutneys – is not difficult and the results are so much better than anything you can buy.

It is also very satisfying to know that you are following in the footsteps of centuries of country housewives who have preserved seasonal foodstuffs to enjoy throughout the year. Before fridges and freezers, fruits, vegetables and herbs were dried, preserved in honey and sugar, or spiced in vinegar. Nuts were stored in wet salt or buried in dry earth, while eggs were pickled, or waxed and buried in sand or sawdust. Scents, flavourings and essences from fruits, nuts, herbs and flowers were distilled and sealed in flasks, while vegetables, flowers, fruits and herbs produced a variety of wines and drinks.

The great country houses had a special 'still-room', where the lady of the house supervised the making of preserves and sweetmeats for the dessert course and the distilling of medicines and perfumes. Several of these early still-rooms survive in National Trust houses: Ham House near Richmond is Surrey is a good example. Their laboratory character gave way to a more conventional kitchen style in the eighteenth century, but the idea that these were places where luxury items were made continued. Usually the still-room was in the housekeeper's part of the house, rather than the cook's. Here jams, jellies, syrups and sauces, and the newly fashionable chutneys and pickles, were made and kept under lock and key. Good examples of nineteenth-century still-rooms can be seen at Uppark and Petworth, both in Sussex, and at Tatton Park in Cheshire.

These arrangements were echoed right down the social scale, so that the humblest cottage might boast a jam cupboard, using the rich harvests of summer to brighten the dull culinary months, and to provide foodstuffs such as pickled damsons, rhubarb jam and quince cheese for the celebration of Christmas and other special occasions.

My family are addicted to making preserves. We all have our own speciality, but agree that it is worth investing in a good-quality stainless-steel preserving pan. A sugar thermometer and a jam funnel are useful, but not essential. All other necessary items are everyday pieces of kitchen equipment.

Although preserving is not difficult, it is important to remember some basic rules about hygiene. Bottles, jars and lids must be clean, free of cracks, and sterilised before use, to prevent the growth of moulds and yeasts. Always pot preserves straight from the pan or totally cold: never when just hot or warm, because this will lead to condensation and mould on the preserve.

To sterilise jars, bottles and lids

Wash in hot, soapy water or the dishwasher, making sure that there is no residue on them, and then rinse thoroughly in hot water. Sterilise using one of the following methods:

- Stand the jars or bottles and lids right-side up on a wire rack in a large pan, making sure that they do not touch each other or the sides of the pan. Cover completely with water and then bring to the boil. Simmer for 10 minutes and then remove from the water and stand upside down on a clean, thick cloth to drain. Dry completely in a pre-heated oven at 110°C, 225°F, gas mark ¼, right-side up, on a baking sheet for about 15 minutes. They can be kept warm in the oven until required.
- Heat the oven to 180°C, 350°F, gas mark 4. Stand the jars, bottles and lids on the oven shelf and leave for 10 minutes to sterilise. Turn the oven off and keep them warm until ready to fill.
- If your dishwasher has a very hot cycle, you can sterilise your jars, bottles and lids in that.

Conversions

Weight	Liquid measure	Length	Temperature
15g (½oz)	30ml (1fl oz)	5mm (¼in)	110°C, 225°F, gas mark ¼
25g (1oz)	100ml (3½fl oz)	1cm (½in)	120°C, 250°F, gas mark ½
40g (1½oz)	150ml (5fl oz or ¼ pint)	1.5cm (⅝in)	140°C, 275°F, gas mark 1
50g (1¾oz)	200ml (7fl oz or ⅓ pint)	2cm (¾in)	150°C, 300°F, gas mark 2
85g (3oz)	250ml (9fl oz)	2.5cm (1in)	160°C, 325°F, gas mark 3
100g (3½oz)	300ml (10fl oz or ½ pint)	5cm (2in)	180°C, 350°F, gas mark 4
175g (6oz)	350ml (12fl oz)		190°C, 375°F, gas mark 5
225g (8oz)	400ml (14fl oz)		200°C, 400°F, gas mark 6
300g (10½oz)	450ml (16fl oz or ¾ pint)		220°C, 425°F, gas mark 7
350g (12oz)	600ml (20fl oz or 1 pint)		230°C, 450°F, gas mark 8
400g (14oz)	850ml (1½ pints)		240°C, 475°F, gas mark 9
450g (1lb)	1 litre (1¾ pints)		
700g (1½lb)	1.2 litres (2 pints)		
900g (2lb)	2 litres (3½ pints)		
1.35kg (3lb)	3 litres (5 pints)		
1.8kg (4lb)	4.2 litres (7 pints)		
2.75kg (6lb)	5 litres (9 pints)		

These approximate conversions are used throughout this book. Please follow either metric or imperial measurements for the correct proportions.

Chutneys, Relishes and Sauces

Nothing livens up simple food like a chutney, relish or ketchup. They are all oriental in origin – chutney coming from the Hindustani word *chatni* meaning a strong, sweet relish and ketchup derived from the Chinese *koe-chiap*, a pickled fish sauce – and first appeared in Britain at the end of the seventeenth century as a result of increased trade through the East India Company.

Chutneys and ketchups continued to exist in their oriental form until the nineteenth century, when the fashion-conscious cooks of British families started to make their own based on the exotic oriental recipes, cooking green peaches, mangoes and bananas with onions, garlic and chillies, and then mixing with smoked fish, spices, olive oil and vinegar. As the taste for chutneys and ketchups developed, the oil was omitted and sugar was added.

The use of home-grown produce, such as plums, apples, and particularly tomatoes, became popular; we could not imagine life today without tomato sauce or ketchup. Generally, chutneys and ketchups became much blander and sweeter and remained so until relatively recently. Now once again we seem to be enjoying spicier and hotter preserves, as part of our love affair with the chilli pepper.

Many National Trust restaurants and tea rooms make their own excellent chutneys and relishes from vegetables and fruit grown in the kitchen gardens of historic houses, on the surrounding estate, or sourced from local farms or market gardens.

A chutney is made by slowly cooking a combination of fruits and vegetables with sugar, vinegar and spices until a jam-like consistency is achieved, whereas the cooking time for a relish is shorter, so the finished preserve has recognizable pieces of the ingredients. The latter can be eaten immediately and should not be kept longer than six weeks, while a chutney is stored for several months, until mellow, before eating and will keep unopened for at least one year.

To Make Chutneys and Relishes

Equipment needed
- Use a stainless-steel or enamel-lined preserving or large pan. Old-fashioned aluminium will react with the vinegar and taint the chutney, as will pans of brass, copper or iron. A shallow pan is best for boiling down chutney because it is quicker, so the preserve tastes fresher.
- A small Pyrex jug makes the best ladle for chutney as the handle slots over the edge of the pan in between filling jars. Invest in a jam funnel, which is cheap and makes it much easier to fill jars.
- Choose heatproof jars with glass lids, or metal lids with a plastic coating on the inside. (Bare metal will react with the chutney). Vinegar will evaporate through cellophane and dehydrate the surface. Make sure the jars are sterilised and dry (see page 7).

Choosing and preparing the vegetables and fruit
- The National Trust recommends sourcing local, seasonal produce where possible, rather than buying imported fruit or fruit that is out of season.
- Vegetables and fruit for chutney can be soft and slightly over-ripe, but must be sound.
- When choosing citrus fruit it is preferable to select unwaxed/organic versions, particularly if you want to use its zest, as many fruits are sprayed with wax to prolong their shelf life.
- Peel and chop finely, mince or process. Onions, garlic and other ingredients, which need long cooking to tenderise them, can initially be cooked separately in water, as sugar and vinegar tend to harden rather than soften them.

Choosing the vinegar and sugar
- Any type of vinegar can be used, depending on the flavour and colour required for the finished chutney, but always use good-quality vinegar.
- Brown sugars give rich colours and heavier flavours than white. If you want a pale-coloured chutney, add the sugar near the end of the cooking time with some of the vinegar. Black treacle and honey make good alternative sweeteners. Dried fruit adds sweetness as well as texture and colour.

Choosing the spices
- Spices can be whole or ground. Ground spice is easier to handle, but whole spice gives the chutney a better flavour.
- When whole spices are used, they should be bruised and tied in a small square of muslin and removed before potting.

Cooking and potting the chutney
1. Cook the vegetables and/or fruit with the sugar, vinegar and spices very slowly, uncovered, until the mixture has thickened and all the excess liquid has evaporated, stirring frequently. This can take from 1–4 hours, but usually 1½ hours is long enough. Generally the longer a chutney is cooked, the more mellow the flavour and the darker the colour. To tell when the chutney is cooked, make a channel right across the surface with a wooden spoon and if this does not fill with vinegar, then it is ready.
2. While still hot, spoon into prepared jars, filling up to the rim, and seal.
3. Label and date the jars. Store chutneys in a cool, dry, dark cupboard, usually for at least one month to mature. Relishes can be eaten immediately and should not be kept longer than six weeks. Once opened, both should be eaten quickly.

To sterilise sauces
It is advisable to sterilise sauces and ketchups after bottling if you want to keep them longer than six months. Otherwise, store in the fridge and use up quickly once opened.

1. Pour the hot sauce through a sterilised funnel into hot, sterilised bottles (old sauce bottles with well-fitting screw caps are ideal) to within 2.5cm (1in) of the top and then lightly screw on the caps.
2. Stand the filled bottles in a large, deep pan and fill with hot water up to their necks. Bring to the boil and then simmer for 30 minutes.
3. Lift out the bottles using an oven cloth and stand them on newspaper to prevent cracking. Tighten the screw caps and leave until cold before storing.

Cliveden Apple Chutney

This is the classic apple chutney recipe made at the National Trust restaurant in the old conservatory at Cliveden, near Maidenhead in Berkshire.

225g (8oz) onions, peeled and finely chopped
700g (1½lb) cooking apples, peeled, cored and roughly chopped
600ml (1 pint) malt vinegar
350g (12oz) soft brown sugar
100g (3½oz) sultanas
15g (½oz) mixed spice
15g (½oz) sea salt
1 teaspoon ground ginger

Cook the onions in boiling water for 5 minutes to soften them, then drain. Place all the other ingredients in a large pan, and add the drained onions. Bring to the boil very slowly, stirring continuously until the sugar has dissolved. Simmer uncovered for 1½–2 hours, until thick, stirring frequently to prevent sticking.

Pour into warm, sterilised jars, seal and store for at least one month before using.

Trerice Apple, Onion and Sage Chutney

This is a light fruity chutney flavoured with sage, which is particularly good with pork and poultry. It is one of several made and sold at the restaurant at Trerice, a small Elizabethan manor house near Newquay in Cornwall owned by the National Trust. They also make a version using mint instead of sage.

1·5kg (3lb 5oz) onions, peeled and finely chopped
1·5kg (3lb 5oz) cooking apples, peeled, cored and chopped
450g (1lb) sultanas
Grated rind and juice of 2 lemons
700g (1½lb) light muscovado sugar
600ml (1 pint) malt vinegar
4 tablespoons chopped fresh sage

Cook the onions in boiling water for 5 minutes to soften them, then drain. Place all the other ingredients, except the sage, in a very large pan, and add the drained onions. Bring to the boil very slowly, stirring frequently to dissolve the sugar. Simmer uncovered for 1½–2 hours, until thick, stirring frequently to prevent sticking.

Stir in the chopped sage, pour into warm, sterilised jars and seal. Leave to mature for two to three months.

VARIATION
Apple and Onion Chutney
Omit the sage.

Mottisfont Fresh Bramley and Date Relish

This uncooked relish or chutney was invented for the tea room at Mottisfont Abbey Garden, a National Trust property near Romsey in Hampshire, where the National Collection of Old-fashioned Roses is held. The chutney has now become popular in other National Trust restaurants in the area. It can be used immediately, but does improve in flavour if kept in the fridge for a few months. Any cooking apples can be used, but Bramleys give a good result.

450g (1lb) Bramley apples, peeled and cored
450g (1lb) sultanas
450g (1lb) onions, peeled
450g (1lb) cooking dates, stoned
450g (1lb) soft brown sugar
600ml (1 pint) malt or white wine vinegar
1 tablespoon sea salt
1 teaspoon ground cloves
1 teaspoon ground allspice
¼ teaspoon cayenne pepper
1 teaspoon ground coriander
50g (1¾oz) fresh root ginger, peeled and finely chopped

Process or mince the apples, sultanas, onions and dates together and place in a large china or glass bowl. Stir in the remaining ingredients, then cover with a clean cloth and leave for 24 hours.

Taste and adjust the seasonings as necessary, then spoon into sterilised jars or into plastic tubs. Store in the fridge.

Windfall Chutney

This recipe is a good way to use windfall apples and is a good all-round chutney, although particularly tasty with grilled cheese or sausages.

900g (2lb) onions, peeled and finely chopped
2.25kg (5lb) cooking apples
60g (2oz) garlic, peeled and finely chopped
60g (2oz) fresh root ginger, peeled and finely chopped
1 large red chilli, deseeded and finely chopped
1·2 litres (2 pints) distilled malt vinegar
500g (1lb 2oz) light muscovado sugar
2 tablespoons ground turmeric
1 tablespoon cooking salt

Cook the onions in boiling water for 5 minutes to soften them, then drain. Cut away any bruised spots from the apples, then peel and core: the prepared apples should weigh around 2kg (4½lb). Chop them finely and place in a very large pan with the onions. Add the garlic, ginger and chilli add to the pan, then pour in the vinegar. Stir in the sugar, turmeric and salt and then bring slowly to the boil, stirring until the sugar has dissolved.

Simmer uncovered for about 1 hour, until thick, stirring frequently to avoid it catching on the bottom of the pan. Spoon into warm, sterilised jars and seal. Store for at least one month before using.

Spiced Blackberry Chutney

This chutney has a delicious fruity flavour. Try to pick larger berries if possible as they are juicier. The chutney is sieved after cooking to remove the pips.

1.35kg (3lb) blackberries
450g (1lb) cooking apples, peeled, cored and chopped
450g (1lb) onions, peeled and finely chopped
1 tablespoon cooking salt
15g (½oz) dry mustard powder
25g (1oz) ground ginger
1 teaspoon ground mace
½ teaspoon cayenne pepper
600ml (1 pint) white wine vinegar
450g (1lb) soft brown sugar

Pick over and wash the blackberries. Put all the ingredients except the sugar in a large pan. Bring to the boil and cook gently for about 1 hour until soft. Push the chutney through a nylon sieve and return to the pan. Stir in the sugar, then heat gently until it has completely dissolved. Bring to the boil, then reduce the heat and simmer gently for about 30 minutes, until thick.

Pour into warm, sterilised jars and cover. Leave to mature for two to three months.

Elderberry Chutney

900g (2lb) elderberries
450g (1lb) cooking apples, peeled, cored and finely chopped
450g (1lb) onions, peeled and chopped
450g (1lb) seedless raisins
1 teaspoon ground cinnamon
1 teaspoon paprika
1 teaspoon ground ginger
¼ teaspoon cayenne pepper
225g (8oz) granulated sugar
300ml (½ pint) distilled malt vinegar

Carefully strip the elderberries from their stalks, then wash well. Cook gently for 10–15 minutes until soft, then push the berries and their juice through a nylon sieve into a bowl to remove the pips. Pour the elderberry pulp into a large pan with all the other ingredients. Bring to the boil very slowly, stirring frequently to dissolve the sugar. Simmer uncovered for 1½–2 hours, until thick, stirring frequently to prevent sticking.

Pour into warm, sterilised jars and seal. Store for at least one month before using.

Mango Chutney

During the nineteenth century, many families spent time in India and the colonies, acquiring an interest in oriental food. It must have been very frustrating when they came home and were unable to get hold of the exotic fruits and vegetables needed for their newly discovered recipes. Mangoes were one of the fruits that needed a substitute. A 'Bengal Recipe for Making Mango Chutney', recommended by Mrs Beeton, uses sour apples instead. Fortunately, mangoes are now widely available, and this popular curry accompaniment can easily be made in your own kitchen. The result will be completely different from the traditional mango chutney made with salted, pickled green mangoes.

1kg (2¼lb) firm mangoes
50g (1¾oz) fresh root ginger, peeled
50g (1¾oz) medium-strength red chilli, deseeded
250ml (9fl oz) cider vinegar
200g (7oz) caster sugar

Dice the flesh of the mangoes, discarding the large, flat stones and the skin. (You can buy a mango-splitting gadget to help you do this, which is well worth acquiring if you are going to make this chutney regularly and I am sure you will).

Cut the ginger and chilli into fine julienne strips. Place all the ingredients in a pan and bring slowly to the boil, stirring continuously until the sugar has completely dissolved. Simmer uncovered for 1 hour.

Spoon into warm, sterilised jars and seal.

Granny's Hot

Given to me by a very special friend in Cornwall, this recipe originating in Ceylon has been passed down in her family for many generations. She recommends it to accompany cold meats and hot grilled mackerel.

450g (1lb) onions, peeled and finely chopped
700g (1½lb) cooking apples, peeled, cored and sliced
450g (1lb) seedless raisins, chopped
4 large garlic cloves, peeled and finely chopped
700g (1½lb) dark soft brown sugar
4 teaspoons cooking salt
2 tablespoons ground ginger
3 tablespoons dry mustard powder
2 tablespoons paprika
1 tablespoon ground coriander
1·5 litres (2¾ pints) malt vinegar

Cook the onions in boiling water for 5 minutes to soften them, then drain. Place all the ingredients in a large pan and bring slowly to the boil, stirring until the sugar has dissolved. Simmer uncovered, over a very low heat, for about 3 hours, stirring frequently as the chutney thickens, to prevent sticking.

When the chutney is thick and pulpy and no excess liquid remains, pour into warm, sterilised jars and seal. Leave to mature for at least three months and as long as 18 months.

Banana and Apple Chutney

The passion for growing tender exotic fruit against all the odds of the English climate meant that much money and effort went into growing pineapples, melons and even bananas. At the National Trust's Petworth House in Sussex, Henry Wyndham, 2nd Lord Leconfield, is said to have built a special glasshouse in the mid-nineteenth century and sent his Head Gardener to Kew Gardens to learn how to grow bananas, in the belief that the fruit tasted much better straight from the tree. Leconfield was disgusted to find that, having spent £3000 on producing it, his first banana tasted exactly like any other. This is a typical nineteenth-century recipe.

450g (1lb) onions, peeled and finely chopped
2kg (4½lb) cooking apples, peeled, cored and finely chopped
12 bananas, peeled and thinly sliced
225g (8oz) seedless raisins
1 tablespoon cooking salt
1 teaspoon dry mustard powder
1 teaspoon ground ginger
1 teaspoon ground cinnamon
1 tablespoon medium curry powder
1·2 litres (2 pints) malt vinegar
450g (1lb) granulated sugar

Cook the onions in boiling water for 5 minutes to soften them, then drain. Place all the ingredients except the vinegar and sugar in a very large pan. Add the onions and half the vinegar, then bring slowly to the boil. Simmer uncovered for about 30 minutes, and then stir in the sugar and the remaining vinegar. Continue to simmer for 1½–2 hours, until thick, stirring frequently to prevent sticking.

Pour into warm, sterilised jars and seal. Leave to mature for at least one month.

Autumn Chutney

A great way of using up windfall pears and apples.

2 medium onions, peeled and finely chopped
450g (1lb) cooking pears, peeled, cored and diced
450g (1lb) cooking apples, peeled, cored and diced
Grated rind and juice of 1 lemon
300ml (½ pint) malt vinegar
½ teaspoon ground cinnamon
Large pinch of ground ginger
Large pinch of ground cloves
225g (8oz) soft brown sugar

Cook the onions in boiling water for 5 minutes to soften them, then drain. Put the pears, apples, onions, lemon rind, vinegar and spices into a large pan, then cook over a low heat for about 20 minutes until the fruits are cooked.

Stir in the sugar and lemon juice and continue to cook over a low heat until the sugar has dissolved, stirring frequently, then bring to the boil and simmer uncovered for about 1 hour, until the mixture thickens.

Spoon into warm, sterilised jars and seal. Store for at least one month before using.

Trerice Pear, Orange and Ginger Chutney

Quinces or apples could be used in this chutney, but it is very good with pears. This is another recipe from the restaurant at Trerice, where the cooks make it with local Cornish varieties of pear from the estate orchard.

1·5kg (3lb 5oz) cooking pears, peeled, cored and roughly chopped
225g (8oz) onions, peeled and finely chopped
225g (8oz) seedless raisins, chopped
50g (1¾oz) preserved stem ginger, finely chopped
350g (12oz) demerara sugar
400ml (14fl oz) white wine vinegar
Grated rind and juice of 2 oranges
15g (½oz) dried root ginger
15g (½oz) whole cloves

Put the pears, onions and raisins in a large pan with all the other ingredients except for the root ginger and cloves. Tie these in a small square of muslin and add to the pan. Slowly bring to the boil, stirring frequently until the sugar has completely dissolved. Simmer uncovered for 1½–2 hours, until thick.

Remove the muslin bag of spices, then pot in warm, sterilised jars and seal. Leave to mature, for six months if possible, before using.

Apricot and Ginger Chutney

100g (3½oz) onions, peeled and finely chopped
225g (8oz) ready-to-eat dried apricots
100g (3½oz) preserved stem ginger
200g (7oz) dark muscovado sugar
600ml (1 pint) malt vinegar
½ teaspoon ground ginger
½ teaspoon ground allspice
Large pinch of sea salt

Cook the onions in boiling water for 5 minutes to soften them, then drain. Finely chop the apricots and cut the stem ginger into fine julienne strips. Place the onions, apricots and ginger in a pan with all the other ingredients and heat gently, stirring continuously until the sugar has dissolved.

Bring to the boil, then turn the heat down and simmer gently for about 1½ hours, stirring frequently towards the end of the cooking time, to stop the chutney sticking to the bottom of the pan and burning.

Spoon into warm, sterlised jars and seal. Store for at least one month before using.

Cragside Curried Fruit Relish

This is served in the Stables Restaurant (formerly the Vickers Restaurant) at Cragside, the imposing Victorian country house built by Sir William, later Lord, Armstrong at Rothbury in Northumberland, now owned by the National Trust.

450g (1lb) dried apricots
450g (1lb) dried peaches
450g (1lb) dates, stoned
450g (1lb) seedless raisins
3–4 whole cloves, crushed
450g (1lb) light muscovado sugar
600ml (1 pint) distilled malt vinegar
600ml (1 pint) water
2 teaspoons cooking salt
2 teaspoons mild curry powder

Chop or mince the dried fruit coarsely. This can be done in a food processor. Place all the ingredients in a large pan and heat gently, stirring continuously until the sugar has dissolved, then bring to the boil. Cover and simmer gently for 15–20 minutes.

Spoon into warm, sterilised jars and seal. Store for at least one month to allow the chutney to mature before using.

Beetroot and Ginger Relish

Excellent with cold meats, coarse terrines, oily fish and goat's cheese. Don't be tempted to use up coarse old beetroot in this recipe as you will not be happy with the result.

40g (1½oz) salted butter
225g (8oz) red onion, peeled and finely chopped
75g (2½oz) granulated sugar
450g (1lb) raw beetroot, peeled and grated
2 teaspoons peeled and grated fresh root ginger
30ml (1fl oz) sherry vinegar
125ml (4fl oz) red wine vinegar
Sea salt and freshly ground black pepper

Heat the butter in a pan and add the onion. Cook over a low heat until very soft, but not browned, then stir in the sugar. Add the beetroot, ginger, sherry and red wine vinegar, then simmer gently uncovered for 30 minutes.

Season to taste, then pot the mixture in warm, sterilised jars and seal. Store for at least one month before using.

Courgette Chutney

A recipe for that glut of summer courgettes, when it rains and they go mad! Particularly good with a well-made mature farmhouse cheese.

1.5kg (3lb 5oz) courgettes
1½ tablespoons cooking salt
450g (1lb) tomatoes, skinned and chopped
225g (8oz) onions, peeled and finely chopped
450g (1lb) sultanas
Grated rind of 2 oranges
900g (2lb) granulated sugar
300ml (½ pint) red wine vinegar
300ml (½ pint) malt vinegar
2 teaspoons ground cinnamon

Chop the courgettes into slices and tip into a colander or sieve. Sprinkle with the salt and leave to stand for about 2 hours, then rinse under cold running water and dry in a clean cloth.

Put all the ingredients into a large pan and bring slowly to the boil, stirring frequently until the sugar has dissolved. Simmer uncovered over a low heat for 1½–2 hours, until thick, stirring from time to time to prevent sticking.

Pour into warm, sterilised jars, seal and store for at least one month before using.

Damson Chutney

A number of interesting damson trees are grown in the orchard at the National Trust's Hardwick Hall in Derbyshire, including a variety called Merryweather, which comes, like the Bramley apple, from the nearby town of Southwell in Nottinghamshire. Although a traditional Northern treat, damsons also grow well in the West Country. Serve this sweet, unctuous chutney, which is more like a fruit sauce, with cold meats, particularly ham and pork, poultry and game, raised pies and cheese.

900g (2lb) damsons
450g (1lb) demerara sugar
1 blade of mace
4 whole cloves
4 black peppercorns
½ teaspoon table salt
300ml (½ pint) malt or wine vinegar

Pick over the damsons, removing any stalks and leaves, then wash them. Place all the ingredients, except the vinegar, in a large pan. Simmer gently for about 15 minutes, stirring to dissolve the sugar. Stir in the vinegar and continue simmering gently for 1 hour.

Rub the mixture through a coarse sieve and return to the pan. Bring to the boil and then simmer gently for about 10 minutes, stirring well so that it does not catch on the bottom of the pan.

Pour into warm, sterilised jars and seal. Store for at least two weeks before using.

Spiced Plum and Lime Chutney

2kg (4½lb) cooking plums, stoned and roughly chopped
2 large onions, peeled and finely chopped
5 garlic cloves, peeled and chopped
2 fresh chillies, deseeded and chopped
Grated rind and juice of 2 limes
Grated rind and juice of 1 lemon
5cm (2in) piece of fresh root ginger, peeled and grated
600ml (1 pint) wine vinegar
500g (1lb 2oz) demerara sugar
2 cinnamon sticks
25g (1oz) allspice berries
1 teaspoon black peppercorns

Place the plums in a large pan with all the other ingredients except the cinnamon, allspice berries and peppercorns. Grind these spices in a spice mill or pestle and mortar until you have a fine powder. Add to the pan and then bring slowly to the boil, stirring regularly until all the sugar has dissolved. Simmer uncovered for 1½–2 hours until thick.

Ladle into warm, sterlised jars and seal. Store for at least two months to mature before eating.

VARIATIONS
Spiced Damson and Lime Chutney
Substitute damsons for plums.

Plum, Lime and Coriander Chutney
Replace the allspice berries with 2 tablespoons coriander seeds.

Fig Relish

This is a recipe invented by my sister Marilyn. It is fantastic with cheese, particularly soft goat's cheese.

225g (8oz) dried figs
½ onion, peeled and chopped
275ml (10fl oz) cider vinegar
100g (3½oz) dark muscovado sugar
25g (1oz) fresh root ginger, peeled and grated
¼ teaspoon yellow mustard seeds
½ small cinnamon stick
Large pinch of ground allspice
Large pinch of ground cloves
½ fresh chilli, including seeds, finely chopped
½ teaspoon table salt

Finely chop the figs and soak overnight in water, to just cover.

Next day, drain the figs. Cook the onion in boiling water for 5 minutes to soften, then drain. Put the figs, onion and all the other ingredients into a pan. Heat gently, stirring continuously until the sugar has completely dissolved, then bring to the boil. Reduce the heat and simmer gently for about 1 hour until the mixture has thickened.

Spoon into warm, sterlised jars and seal. Keep in the fridge. Ready to eat immediately.

Gooseberry Chutney

Make this recipe when gooseberries are cheap – unripe green fruit is best. It is delicious with fresh mackerel, smoked fish and roast poultry, as well as cheese.

1kg (2½lb) gooseberries, topped and tailed
300ml (½ pint) malt vinegar or white wine vinegar
100g (3½oz) seedless raisins
100g (3½oz) sultanas
1 large onion, finely chopped
450g (1lb) soft brown sugar
Pinch of ground turmeric
1 teaspoon freshly ground black pepper
1 teaspoon ground mixed spice
1 teaspoon ground cinnamon

Wash the gooseberries and place in a large pan. Add the vinegar and simmer gently for about 20 minutes, or until the fruit is soft and pulpy. Add all the other ingredients and continue simmering, stirring frequently to dissolve the sugar, for about 1½–2 hours until thick.

Pour into warm, sterilised jars and seal. Leave to mature for two to three months.

Marrow, Red Tomato and Date Chutney

This is an inexpensive recipe for using up a harvest festival marrow or a glut of tomatoes, especially good with hamburgers and hot dogs.

1·5kg (3lb 5oz) marrow
75g (2½oz) cooking salt
900g (2lb) red tomatoes
225g (8oz) onions
350g (12oz) cooking apples
600ml (1 pint) distilled malt vinegar
225g (8oz) cooking dates, stoned and chopped
450g (1lb) soft light brown sugar
1 tablespoon mustard seeds
2 tablespoons ground ginger
2 teaspoons ground allspice
Large pinch of freshly grated nutmeg

Peel the marrow and cut it in half lengthways. Scoop out the seeds and discard them, then cut the flesh into 1cm (½in) cubes. Put these into a bowl, sprinkling each layer with salt. Cover the bowl with a clean cloth and leave for 24 hours.

The next day, rinse the salted marrow under cold running water, drain well and set aside. Skin the tomatoes by covering with boiling water for a few minutes, then slipping off the skins. Chop the tomatoes, peel and chop the onions and peel, core and slice the apples. Put the tomatoes, onions and apples in a large preserving pan. Add the vinegar, stir well and then bring to the boil over a moderate heat. Reduce the heat and cook gently for 30 minutes. Add the dates, sugar and spices, and the reserved marrow. Stir well and bring to the boil over a moderate heat.

Reduce the heat and simmer gently for 1½–2 hours, stirring frequently as the mixture thickens, to stop it catching on the bottom of the pan.

Pour into warm, sterilised jars and seal. Leave to mature for two to three months.

VARIATIONS
Courgette, Pumpkin or Squash Chutney
Substitute pumpkin or squash for the marrow and prepare in the same way. Courgettes will need only to be trimmed before being cut into pieces.

Fresh Marrow Relish

Ready to eat immediately with cold meat or cheese.

1 medium-sized marrow
1 tablespoon table salt
4 ripe tomatoes
4 thin-skinned lemons
A little sunflower oil
1 red onion, peeled and finely sliced
1 tablespoon granulated sugar
1 tablespoon cider vinegar
½ teaspoon freshly ground white pepper
½ cinnamon stick
1 tablespoon sultanas

Peel the marrow, cut it in half lengthways and remove the seeds. Cut it lengthways again and then into 1cm (½in) slices. Toss them with the salt in a bowl and leave for 1 hour.

Skin the tomatoes by covering with boiling water for a few minutes, then slipping off the skins. Cut them into small dice. Cut the rind and pith from the lemons and slice them into rounds, discarding the pips.

Heat a little oil in a large pan and cook the onion gently until soft but not browned. Briefly rinse the marrow under cold running water. Dry on kitchen paper then add to the onion. Turn up the heat, add the sugar and cook the mixture for 10 minutes, stirring frequently to avoid colouring. Add the vinegar and reduce to a syrupy glaze.

Add the tomatoes, lemon slices, pepper and cinnamon, then cover the pan and simmer very gently for about 20 minutes. Add the sultanas and cool before serving.

Pumpkin and Raisin Chutney

An excellent way of using up the flesh after making a pumpkin lantern for Hallowe'en. Any variety of squash can be used in the same way.

1kg (2¼lb) pumpkin flesh, chopped
900g (2lb) brown or white granulated sugar
225g (8oz) seedless raisins
600ml (1 pint) distilled malt vinegar
100g (3½oz) onion, peeled and chopped
¼ teaspoon freshly grated nutmeg
25g (1oz) cooking salt
1 teaspoon ground ginger
1 teaspoon freshly ground black pepper
Juice of 1 lemon
2 tablespoons grape juice
4 bay leaves

Place the pumpkin flesh in a large bowl with all the other ingredients. Stir well, then cover with a clean cloth and leave for about 3 hours.

Transfer to a large pan and then bring slowly to the boil, stirring frequently to dissolve the sugar completely. Simmer uncovered for 1½–2 hours, until thick.

Remove the bay leaves, then pot into warm, sterilised jars and seal. Store for about one month before using.

Red Onion Marmalade

Another seriously yummy recipe from my sister. So many onion marmalades on the market are disappointing, but not this one, which is full of good things, including red wine and port. You need to make some Sweet Pickling Vinegar before you start (see page 62).

My sister recommends eating this marmalade with cheese scones, fresh bread and cheese, and with burgers; and I like to use it to baste sausages or lamb chops, a few minutes before cooking is completed.

900g (2lb) red onions
2 large garlic cloves
2 dessertspoons olive oil
100g (3½oz) dark brown sugar
1 teaspoon ground allspice
Sea salt and freshly ground black pepper
100ml (3½fl oz) red wine
200ml (7fl oz) port
350ml (12fl oz) Sweet Pickling Vinegar (see page 62), made with white wine vinegar
1 tablespoon lemon juice
2 tablespoons redcurrant jelly

Peel and chop, or finely slice, the onions and garlic. Heat the olive oil in a pan and add the onion mixture. Stir until glossy with oil and heated through and then sprinkle over the sugar and allspice. Season with salt and pepper and stir well. Cook, uncovered, over a very gentle heat, stirring occasionally, for about 1 hour. Bring to the boil and continue boiling for 5 minutes to evaporate the juices and develop a rich mahogany colour. (The onions should be soft and break up when pressed between your fingers and the sugar should smell of caramel.)

Pour in the wine, port, Sweet Pickling Vinegar and lemon juice and stir in the redcurrant jelly. Bring to the boil and cook over a high heat for about 45 minutes, stirring frequently towards the end of the cooking time, to stop the mixture sticking and burning on the bottom of the pan. When cooked, the mixture should be thick and rich in colour.

Pot in warm, sterilised jars and seal. Store in the fridge. Ready to eat immediately.

Seville Orange and Apricot Chutney

An excellent relish to serve with sausages, cold meats, pork or game pie, roast pork or a plain roast chicken. Seville oranges are only available for about six weeks in January and early February, although you can freeze them whole or cut into chunks for later use.

2 large onions, peeled and chopped
4 Seville oranges
350g (12oz) ready-to-eat dried
 apricots, roughly chopped
75g (2¾oz) seedless raisins
1 red chilli, deseeded and finely
 chopped
2·5cm (1in) piece of fresh root ginger,
 peeled and finely chopped
6 whole cloves
1 teaspoon freshly grated nutmeg
1 tablespoon black peppercorns,
 roughly crushed
2 teaspoons table salt
600ml (1 pint) distilled malt vinegar
225g (8oz) light muscovado sugar

Cook the onions in boiling water for 5 minutes to soften them, then drain. Scrub the oranges well and remove the zest. Finely chop the zest and then cut the flesh into chunks, discarding the pips.

Place all the ingredients in a large pan and leave to infuse for 30 minutes, and then bring the mixture slowly to the boil, stirring frequently until all the sugar has dissolved. Reduce the heat and simmer uncovered very gently for about 1 hour, or until most of the liquid has evaporated, stirring regularly to prevent sticking.

When the chutney is thick and dark, spoon into warm, sterilised jars and seal. Store in a cool dark place for a week or two, or better still, one month before using.

Red Pepper, Tomato and Lemongrass Chutney

4 tablespoons olive oil
225g (8oz) red onion, finely chopped
450g (1lb) very ripe red peppers, deseeded and finely diced
½ teaspoon table salt
½ teaspoon ground allspice
½ teaspoon ground mace
½ teaspoon freshly grated nutmeg
½ teaspoon grated fresh root ginger
1 stick of lemongrass, finely chopped
450g (1lb) very ripe tomatoes
100g (3½oz) seedless raisins
1 garlic clove, peeled and chopped
225g (8oz) granulated sugar
150ml (¼ pint) white wine vinegar

Heat the oil in a large pan, add the onion and cook gently for 5 minutes without browning. Add the peppers to the pan with the salt, spices, ginger and lemongrass. Cook for 10 minutes over a low heat, stirring occasionally.

Skin the tomatoes by covering with boiling water for a few minutes, then slipping off the skins. Chop and add to the pan with the raisins, garlic, sugar and vinegar.

Bring to the boil slowly, stirring frequently to dissolve the sugar, then simmer very gently for about 1¼ hours, until thick. Ladle into warm, sterilised jars and seal.

Store in a cool, dark place for at least one month before using.

Rhubarb and Coriander Chutney

George Wyndham, 3rd Earl of Egremont, was among the earliest cultivators of rhubarb as a garden plant at Petworth in Sussex in the eighteenth century. It took another hundred years before rhubarb became established as the carefully nurtured plant of Victorian kitchen gardens, where it was forced under tall terracotta pots. This chutney recipe is invaluable for making inroads into a late-season glut of rhubarb that is no longer delicate enough for desserts. Serve with curries and spicy dishes, or with cold meats and cheese.

450g (1lb) onions, peeled and finely chopped
900g (2lb) trimmed rhubarb, cut into short lengths
400ml (14fl oz) red wine vinegar or raspberry vinegar
1½ teaspoons table salt
225g (8oz) seedless raisins
225g (8oz) sultanas
2 teaspoons medium curry powder
175g (6oz) granulated sugar
15g (½oz) whole coriander seeds

Cook the onions in a little boiling water until tender, then drain. Place the rhubarb in a pan with all the other ingredients except the coriander seeds. Lightly bruise these, then tie in a small square of muslin and add to the pan. Gently bring to the boil, stirring frequently until the sugar has completely dissolved, then simmer uncovered for 1½–2 hours, until thick.

Remove the muslin bag of spices and discard. Pour into warm sterilised jars and seal. Leave to mature for at least one month in a cool, dark place before eating.

Wimpole Spring Rhubarb, Melon and Ginger Relish

The kitchen garden at Wimpole Hall, a National Trust property near Cambridge, grows the rhubarb for this unusual chutney, which was invented by one of their restaurant chefs, to serve with locally smoked fish and poultry.

4 garlic cloves, peeled and crushed with a little table salt
100g (3½oz) red onion, peeled and finely chopped
600ml (1 pint) distilled malt vinegar
900g (2lb) forced rhubarb, washed, trimmed and cut into 2.5cm (1in) lengths
1 medium Honeydew melon, peeled, deseeded and cut into 1cm (½in) cubes
175g (6oz) preserved stem ginger, sliced
100g (3½oz) soft brown sugar
150ml (¼ pint) preserved stem ginger syrup

Boil the garlic and onion with the vinegar until it has reduced by half. In a separate pan, simmer the rhubarb, melon and ginger with the sugar and ginger syrup until the rhubarb is tender, but not broken up.

Strain the fruit through a sieve, reserving the juice. Add the juice to the vinegar and boil together until reduced to a syrup (taking care not to burn). Add the fruits to this syrup and bring back to the boil.

Spoon into warm, sterilised jars and seal. Store for at least one month before using.

Rhubarb Sauce

This is superb with fresh and smoked mackerel and roast goose.

225g (8oz) shallots, peeled and finely chopped
1·5kg (3lb 5oz) rhubarb, washed, trimmed and cut into chunks
450ml (¾ pint) malt vinegar
Large pinch of cayenne pepper
5 whole cloves
Small pinch of sea salt
Freshly ground black pepper
450g (1lb) granulated sugar
3 teaspoons ground turmeric
2 teaspoons dry mustard powder

Cook the shallots in boiling water for 5 minutes to soften them, then drain. Place the rhubarb in a large pan with 150ml (¼ pint) of the vinegar, the shallots, cayenne and cloves, salt and a generous grind of black pepper. Bring slowly to the boil, then simmer gently for 1 hour.

Carefully remove and discard all the cloves, then purée the mixture in a blender or food processor. Return the purée to a clean pan, stir in the sugar and the remaining vinegar and bring slowly to the boil, stirring frequently to dissolve the sugar. Mix the turmeric and mustard together and stir into the rhubarb mixture. Reduce the heat and simmer gently for about 30 minutes until nice and thick, but still pourable.

Remove from the heat and allow to cool for 10–15 minutes. Spoon into sterilised jars or bottles and seal. After bottling, the sauce should be sterilised (see page 11) unless you intend to use it up very quickly.

Red Tomato and Ginger Chutney

Another of my family's favourite chutneys, which is fantastic with crumbly cheese, cheese on toast, sausages and cold meats.

4 large onions, peeled and finely chopped
1·5kg (3lb 5oz) tomatoes, skinned and chopped
450g (1lb) cooking apples, peeled, cored and finely chopped
450g (1lb) carrots, peeled and finely chopped
4 teaspoons peeled and grated fresh root ginger
4 garlic cloves, peeled and crushed
225g (8oz) sultanas
2 teaspoons table salt
2 teaspoons mixed spice
850ml (1½ pints) malt vinegar
800g (1lb 12oz) demerara sugar

Cook the onions in a little boiling water until tender, then drain well. Place in a large pan with all the other ingredients except the sugar, then bring to the boil. Reduce the heat and simmer for about 30 minutes until the carrots are tender. Add the sugar and continue cooking gently, stirring continuously until it has dissolved completely. Then simmer for about 1½ hours until thick.

Ladle into warm, sterlised jars and seal. Leave to mature for two to three months.

Father's Special Fruit Chutney

My father invented this chutney because he couldn't find a sweet fruit recipe that he thought was really good and could be made at any time of the year. Well, here it is, and all our family love it, especially at Christmas – hope you do.

225g (8oz) red or green tomatoes
300ml (½ pint) malt vinegar
1kg (2¼lb) cooking apples
225g (8oz) onions, peeled and quartered
450g (1lb) dried apricots
450g (1lb) stoned dates
450g (1lb) seedless raisins
100g (3½oz) preserved stem ginger
2 large garlic cloves, peeled
3g (⅛oz) dried chillies
700g (1½lb) soft brown sugar
50g (1¾oz) cooking salt
1½ tablespoons ground mixed spice

If using red tomatoes, skin them by covering with boiling water for a few minutes, then slipping off the skins. Cut the tomatoes into small pieces and place in a large pan with the vinegar. Cook very slowly for about 30 minutes until soft. Peel, core and cut the apples into quarters: the prepared apples should weigh 900g (2lb).

Mince or process the apples, onions, apricots, dates, raisins, ginger, garlic and chilli and add to the pan. Stir in the sugar, salt and spice then heat gently, stirring well to dissolve the sugar completely. Increase the heat and bring to the boil, then reduce the heat and simmer for about 1 hour until the chutney is thick and rich in colour, stirring frequently to prevent sticking.

Pour into warm, sterilised jars and seal. Store in a cool, dark place for at least one month, or for several months if you want the chutney to be really mature.

Cumberland Sauce

The food writer Elizabeth David rated Cumberland Sauce 'best of all sauces for cold meat and ham, pressed beef, tongue, venison, boar's head or pork brawn'. We also love it with all kinds of roast game as well as raised pies – a must on your Christmas table. Always use a good-quality redcurrant jelly, preferably homemade (see page 140), to get the best-flavoured sauce.

4–6 shallots, peeled and finely chopped
3 thin-skinned oranges
3 lemons
450g (1lb) redcurrant jelly
150ml (¼ pint) ruby port
A little sea salt
1 dessertspoon dry mustard powder
Large pinch of ground ginger
3 tablespoons cider vinegar

Cook the shallots in boiling water for 5 minutes to soften them, then drain. Using a potato peeler, remove the rind from the oranges and lemons. (Make sure that no white pith is taken off with the rind, as this is bitter.) With a very sharp knife, shred the rind as finely as you can.

Put the shredded peel into a small pan and pour over enough water to cover. Bring to the boil, then immediately pour into a strainer. Cool the peel under cold running water for a minute or two and then put on one side.

Squeeze and strain the juice of 2 of the oranges and 2 of the lemons. Add the shallots and all the other ingredients except for the reserved peel. Heat very gently, stirring all the time, until the redcurrant jelly is melted, then bring to the boil and simmer uncovered over a low heat for 15 minutes, stirring frequently to ensure that the sauce does not catch. Add the shredded peel and simmer for a further 5–10 minutes until the sauce starts to thicken.

Pour into warm, sterilised jars or bottles and seal. Cool, then refrigerate until fully thickened. Serve chilled.

Mushroom Ketchup

An eighteenth-century mushroom ketchup was usually no more than a concentrated mushroom essence and salt was used as the preservative. The product made commercially in the nineteenth century was more elaborate, and it combined mushrooms with anchovies, shallots and spices, but still with quantities of salt. Today, vinegar is used as the main preservative and the ketchup is a useful flavouring for soups, stews and casseroles.

The best mushrooms for making ketchup are large, black and flat, as these give a good quantity of highly flavoured juice.

1·5kg (3lb 5oz) large mushrooms
2 tablespoons sea salt
1 litre (1¾ pints) cider vinegar or wine vinegar
10 black peppercorns
3 blades of mace
½ teaspoon ground ginger
¼ teaspoon freshly grated nutmeg

Wipe the mushrooms clean with kitchen paper and cut off the base of the stalks. Break them into small pieces and layer in a bowl, sprinkling with the salt. Cover with a clean cloth and leave overnight.

Next day, crush the mushrooms with a wooden spoon and tip into a pan with the vinegar and spices. Bring slowly to the boil, stir and then reduce the heat. Cover the pan and simmer for 30 minutes.

Rub the mixture through a fine sieve and pour into warm, sterilised bottles. Seal and then sterilise the ketchup (see page 11).

Tomato Ketchup

If you have a glut of cherry tomatoes in the garden or greenhouse, have a go at making your own tomato sauce. It will be completely different from Heinz, but tastes great and is such a treat with a steak, sausages or burgers. The redder the tomatoes, the better the colour of the sauce.

- 1 large red onion, peeled and roughly chopped
- 1 stick of celery, roughly chopped
- 2·5cm (1in) piece of fresh root ginger, peeled
- 2 large garlic cloves, peeled and chopped
- ½ fresh red chilli, deseeded and finely chopped
- 1 teaspoon ground allspice
- 1 tablespoon coriander seeds
- Pinch of cayenne pepper
- 1 teaspoon freshly ground black pepper
- Pinch of sea salt
- 500g (1lb 2oz) cherry tomatoes
- 400g (14oz) tinned plum tomatoes
- 350ml (12fl oz) cold water
- 200ml (7fl oz) red wine vinegar
- 60g (2oz) light muscovado sugar

Put the onion and celery in a large pan with the ginger, garlic, chilli, allspice and coriander seeds. Season with cayenne, black pepper and salt. Cook gently over a low heat for 10–15 minutes until the vegetables are soft, stirring now and again. Add all the tomatoes and the water and then bring gently to the boil. Simmer gently uncovered until the mixture has reduced by half.

Purée in a blender or food processor and then push through a fine sieve to produce a smooth, shiny sauce. Pour this into a clean pan and add the vinegar and sugar. Simmer gently, stirring frequently until the sugar has dissolved, and continue simmering until the mixture begins to thicken. Taste and adjust the seasoning as necessary.

Pour into warm, sterilised bottles and seal. Store in the fridge, where it will keep for six months; or sterilise the ketchup (see page 11).

Pickles

Pickling is one of the most ancient methods of preserving fruit and vegetables. The Romans were very keen on pickling to preserve a part of their fruit and vegetable harvest; they used vinegar from wine that had gone flat. Wine vinegar remained popular for pickles and chutneys right up to the nineteenth century, which is why pickles used to have a milder flavour than our modern ones made with malt vinegar. We are now beginning to rediscover the taste for pickling with wine vinegar.

Vines were re-introduced to England in medieval times by monks, and the grapes which remained unripe at the end of the season were fermented to form 'verjuice', a kind of sharp vinegar used in pickling. Where grapes were not available, crab apples were also made into verjuice. The variety of pickles was limited by the produce available. As the variety increased in the 16th and 17th centuries, so pickling grew in popularity. Pickled vegetables, herbs, walnuts, flowers and fruit added colour and flavour to fresh salads, meat and fish dishes, and also supplied winter salads. 'Piccalilli', an elaborate Indian pickle still popular today, was adopted by English cooks as early as 1694. Pieces of cabbage, celery and other vegetables were placed in a brine and vinegar sauce, flavoured with ginger, garlic, pepper, bruised mustard seeds and powdered turmeric.

In the latter part of the seventeenth century, the British East India Company imported new fruits and the variety of pickles grew ever larger. Pickled mangoes were copied using home-grown melons, cucumbers, peaches, onions or plums. By the end of the nineteenth century most of today's pickle manufacturers were established. A number of the old recipes have disappeared, but many National Trust restaurants are now making their own pickles and hope to re-introduce some that have been long-forgotten.

To Make Pickles

Equipment needed
- Stainless-steel or enamel-coated iron preserving or large pan.
- Heatproof, wide-mouthed jars with vinegar-proof, airtight covers. Vinegar corrodes bare metal lids and this will taint and discolour your pickle, so make sure any metal lids have plastic-coated linings, or use plastic covers. If your lids are not airtight, the vinegar will evaporate.
- A large china or glass bowl, to hold the vegetables in brine.

Choosing and preparing the vegetables and fruit
- Use only young fresh vegetables or perfect, just-ripe fruit.
- When choosing citrus fruit it is preferable to select unwaxed/organic versions, particularly if you want to use its zest, as many fruits are sprayed with wax to prolong their shelf life.
- Cut into suitably sized pieces if necessary.
- Vegetables are usually soaked in brine or layered with dry cooking salt before bottling. This draws out moisture from the tissues that could otherwise dilute the vinegar and reduces its preservative quality. For a very crisp pickle, dry salting is best. Rinse off excess salt from the vegetables and drain thoroughly.
- Fruit is usually cooked in sweetened, spiced vinegar before bottling. Prick the skins of whole fruit before cooking or they may shrivel.
- Pack the prepared vegetables or fruit into dry sterilised jars. Drain any excess water in the jar.

Completing the pickle
1. Choose a good-quality vinegar and one that suits the particular pickle you are making. For example, if you want to preserve the colour of a certain vegetable, such as red cabbage, use a white vinegar, whereas eggs are better pickled with the stronger-flavoured malt vinegar.
2. Flavour your vinegar with herbs and spices. It is a good idea to have a ready supply of Spiced Vinegar (see page 60) rather than having to make it up at the last minute.
3. Pour the spiced vinegar or syrup over the pickles in the jars to cover them by at least 1cm (½in). The vegetables or fruit tend to absorb the liquid and those at the top may become exposed, so you may have to top up the vinegar after a few days. Use cold vinegar unless a soft pickle is preferred, for example for walnuts or damsons, when boiling vinegar should be added, making sure the jars are warm.

Pickled Chillies

Chillies are fun and easy to grow. You will find you always grow too many to eat fresh, so try this way of preserving them for future use. They are excellent with cheese and bread for lunch, or as an accompaniment to a burger. Remember to wear rubber gloves when preparing chillies and don't touch your eyes. Leave the seeds in if you like extra heat.

12 mild or medium fresh chillies
2 teaspoons granulated sugar
2 teaspoons whole mustard seeds
1 teaspoon table salt
About 150ml (¼ pint) white wine vinegar
2 tablespoons water

Wash the chillies and slice them into thick rings, removing the seeds if you wish. Pack tightly into warm, sterilised jars. Place the remaining ingredients in a small pan and bring to the boil. Pour over the chilli rings to cover, then seal the jars.

Store in a cool dark place for at least one month before using. The longer you leave them, the hotter they get; they will keep for a year.

Pickled Baby Beets

The very small beetroot (the thinnings from the crop) are sweet and delicious. Pickle them whole or slice them thinly. They are very good flavoured with horseradish, but any spiced vinegar will do. Serve with cold roast beef.

Baby beets
A few juniper berries, crushed
Horseradish Vinegar (see below)
Cooking salt

Heat the oven to 180°C, 350°F, gas mark 4. Wash the beets carefully, taking care not to damage the skins or they will lose their lovely colour. Wrap in foil and bake for about 1 hour until tender. Cool, then peel them. Pack them in a wide-mouthed jar with a few crushed juniper berries.

For every 600ml (1 pint) vinegar add 15g (½oz) cooking salt, stirring until it dissolves. Pour the vinegar into the jar to cover the beets. Seal and leave to mature for about one week; the pickles will keep for a year without deteriorating.

Horseradish Vinegar

It is an awful job to prepare fresh horseradish, but it does have a much better flavour than dried. Wear rubber gloves and hold the horseradish under water while scrubbing and peeling it, and preferably work in the open air to minimise the eye-burning effect. Use this vinegar for salad dressings and for pickling.

75g (2¾oz) fresh horseradish, grated
1 shallot, peeled and finely chopped
1 tablespoon mixed whole peppercorns
½ teaspoon cayenne pepper
1·2 litres (2 pints) distilled malt vinegar

Put all the ingredients into a bowl or plastic container, cover and leave to stand for one week. Strain into clean, dry bottles.

Spiced Blackberries

These spiced fruits are extremely good with cold meats, poultry or soft cream cheeses.

300ml (½ pint) Spiced Vinegar (see below) made with white wine vinegar
450g (1lb) granulated sugar
1·35kg (3lb) large clean blackberries
3 rose geranium leaves, washed and dried (if available)

Gently heat the vinegar with the sugar until the sugar has completely dissolved. Simmer for 5 minutes and then add the blackberries. Simmer for a further 5–6 minutes until the blackberries are soft, but not disintegrating. Remove the fruit with a draining spoon and pack into warm, sterilised jars. Boil the vinegar and sugar rapidly for about 5 minutes until reduced to a thick syrup. Add the geranium leaves to the fruit in the jars and pour in the hot syrup. Seal and store for two to three weeks before using.

Spiced Vinegar

This spiced vinegar is used in a large number of pickles and chutneys. It must be made at least six to eight weeks before you intend using it, as the spices need to steep in the vinegar for maximum flavour. Whole spices give a better result.

1 tablespoon celery seeds
1 tablespoon whole mustard seeds
1 tablespoon green cardamom pods, bruised
2 teaspoons coriander seeds
2 teaspoons whole cloves
2 teaspoons allspice berries
10 dried red chillies
1 tablespoon black peppercorns, lightly crushed
25g (1oz) fresh root ginger, peeled and finely sliced
1·2 litres (2 pints) cider, malt or white wine vinegar

Mix all the spices together in a bowl, then divide them between two clean, dry bottles and fill to the top with vinegar. Cover and leave to stand for six to eight weeks, shaking occasionally. Strain the vinegar before using.

Crab Apple Pickle

This was probably the first fruit pickle ever made in Britain. It certainly dates back to Roman times and may go back even further as crab apples were one of the first fruits eaten by prehistoric man. We tend to ignore them nowadays because of their sharp taste, but they are still a very useful fruit for pickling and jellying. Serve with cold ham, pork, venison, duck or goose.

1·35kg (3lb) crab apples
Pared rind of ½ lemon
600ml (1 pint) Sweet Pickling Vinegar (see below) made with white wine vinegar

Choose crab apples of even size, if possible, and wash them. Remove any stalks and dry the apples thoroughly. Prick all over with a darning needle or skewer. Put the pared lemon rind and vinegar into a pan and bring to the boil. Add the apples and cook gently, until almost tender.

Use a draining spoon to remove the fruit and pack carefully into hot, sterilised jars. Discard the lemon rind from the vinegar, then boil rapidly for 5 minutes. Pour over the apples and seal at once. Store for four to six weeks before using.

Sweet Pickling Vinegar

This vinegar is ideal for sweet pickles and fruit pickles. Any vinegar can be used, but use white granulated sugar with white vinegar to preserve its colour.

900g (2lb) brown or white granulated sugar
1·2 litres (2 pints) white distilled, white wine or malt vinegar
1 tablespoon whole cloves
1 cinnamon stick
1 tablespoon coriander seeds
1 tablespoon allspice berries
6 blades of mace

Dissolve the sugar in the vinegar and pour into a large jar or bottle. Put the spices in a muslin bag and add to the vinegar. Cover and leave to steep for six to eight weeks. Strain before using.

Pickled Red Cabbage

This was a popular pickle in Tudor and Stuart times because of its glorious colour that looked so attractive in salads. Turnips and beetroot were often added to make 'a pretty pickle'. Use young firm cabbage and prepare in small batches, as it does not remain crisp for long.

450g (1lb) red cabbage
1 tablespoon cooking salt
½ tablespoon granulated sugar (optional)
300ml (½ pint) Spiced Vinegar (see page 60) made with white wine or cider vinegar
Several slices of raw beetroot
Several slices of raw onion

Quarter the cabbage, removing the outer leaves and centre stalks, as only the heart will make a good pickle. Shred finely and layer it with the salt in a large flat dish or bowl. Cover with a clean cloth and leave to stand overnight.

Dissolve the sugar (if using) by gently heating it in the vinegar, then leave until cold. Rinse the salt off the cabbage and drain thoroughly in a sieve. Pack into clean, dry jars and top with a slice of beetroot, for colour, and a slice of onion, for flavour. Pour in the cold vinegar, then cover and seal. Leave for one week before eating, but use within two months or it will lose its colour and crispness.

Variations
Pickled White Cabbage
Use white cabbage instead of red.

Pickled Red Cabbage and Carrots
Wash 225g (8oz) carrots, peel and cut them into matchstick pieces. Layer the carrot and cabbage in the jars and continue as above.

Pickled Red Cabbage with Oranges and Raisins
Add the grated rind of 2 oranges, the flesh of the oranges cut into segments and 50g (1¾oz) raisins to the vinegar. Bring to the boil, then leave until cold. Continue as above.

Sweet Damson Pickle

If you wish, you can save time by using already prepared Sweet Pickling Vinegar (see page 62), made with white wine or cider vinegar. Plums or greengages can be pickled in the same way, but store for three months before using. Serve with poultry, particularly duck and goose, game or cold meat.

450g (1lb) large damsons
300ml (½ pint) white wine or cider vinegar
225g (8oz) granulated sugar
Pared rind of ½ lemon
2 pieces of dried root ginger, bruised
1 blade of mace
4 whole cloves
1 cinnamon stick
Blackcurrant leaves

Prick the damsons all over with a darning needle and place in a pan. Cover with vinegar and add the sugar. Tie the lemon rind, bruised ginger and other spices in a small square of muslin and add to the pan. Cook very gently until the sugar has dissolved, then bring to the boil. Continue to cook very gently until the damsons are just tender, but do not let the skins break.

Place a blackcurrant leaf in the bottom of each warm, sterilised jar, then use a draining spoon to transfer the fruit carefully to the jars. Discard the muslin bag and boil the vinegar rapidly for about 5 minutes until reduced to a syrup. Pour immediately over the damsons and seal the jars. They are ready to eat in four weeks, but are much better if kept longer.

Indian Pickle

Indian Pickle, or Piccalilli, was one of the first pickles to be brought back by the British East India Company at the end of the seventeenth century. English cooks adopted the recipe with enthusiasm. A 1765 example from Erddig, a National Trust property near Wrexham, reads 'Take vinegar one gallon, garlic one pound, ginger one pound, turmeric, mustard seed, long pepper and salt of each 4 ounces'. The garlic had to be peeled and salted for three days, then washed and salted again and left for another three days, then washed and dried in the sun. All the vegetables to be put into this sauce had to be cut into walnut-sized pieces.

A variety of seasonal vegetables can be used, but cauliflower and baby onions should always be included to give the anticipated crunch. Enjoy with cheese or cold meats.

1·2kg (2lb 10oz) mixed vegetables, cut into 1cm (½in) cubes, with cauliflower cut into small florets and baby onions left whole
2 tablespoons cooking salt
50g (1¾oz) dry mustard powder
2 teaspoons ground ginger
½ teaspoon ground white pepper
¼ teaspoon freshly grated nutmeg
1 tablespoon ground turmeric
50g (1¾oz) plain flour
250ml (9fl oz) cider vinegar
250ml (9fl oz) distilled malt vinegar

Place the prepared vegetables in a large bowl, add the salt and mix well. Cover and leave overnight so that the salt draws out the liquid, giving the vegetables extra bite and flavour.

Next day, drain and rinse the vegetables under cold running water, then dry well.

Make the sauce by mixing the spices and the flour with enough cider vinegar to make a smooth paste. Whisk in the remaining cider and malt vinegar and pour into a large pan. Bring slowly to the boil, whisking continuously to give a smooth, thick sauce. Simmer for 4–5 minutes, then add the vegetables. Bring back to the boil and simmer for about 3 minutes. Be careful not to overcook as you want the vegetables to be crunchy.

Spoon into warm, sterilised jars, seal and store. Eat within three months.

Cotehele Sweet Piccalilli

The restaurant at Cotehele, the National Trust's romantic medieval manor house on the banks of the River Tamar in Cornwall, serves this sweet piccalilli with a platter of West Country cheeses and home-baked bread. The delicious sauce in the pickle includes curry powder.

1·2 litres (2 pints) water
75g (2¾oz) cooking salt
450g (1lb) small pickling onions or shallots, peeled
700g (1½lb) small cauliflower florets
600g (1lb 5oz) cucumber, cut into 1cm (½in) dice
450g (1lb) fine green beans, trimmed and halved
3 small fresh red chillies, deseeded and cut into thin strips

FOR THE SAUCE
225g (8oz) granulated sugar
75g (2¾oz) plain flour
2 teaspoons ground allspice
2 tablespoons ground ginger
2 tablespoons mild curry powder
2 teaspoons ground turmeric
2 tablespoons mustard powder
½ teaspoon cayenne pepper
1·2 litres (2 pints) distilled malt vinegar
25g (1oz) black peppercorns

Bring the water and salt to the boil in a large pan. Add all the prepared vegetables and cook gently for 5 minutes. Pour into a colander, rinse well under cold water, and then dry in a clean cloth.

To make the sauce, put the sugar, flour, allspice, ginger, curry powder, turmeric, mustard and cayenne into a small basin. Add 3–4 tablespoons of the vinegar and mix to a thick paste. Put this paste with the remaining vinegar and the peppercorns into a pan and bring to the boil, stirring continuously. Reduce the heat and continue to cook, stirring continuously, for 3–5 minutes until the sauce thickens. Remove the pan from the heat and leave to cool, stirring occasionally to prevent a skin from forming. Put the drained vegetables into a bowl, add the sauce and mix together. Cover the bowl and leave to stand for 24 hours.

Next day, stir the piccalilli again to coat the vegetables evenly, then spoon into clean, dry jars and seal. Store for two to three months before eating.

Sweet Cucumber Pickle

Good with crusty bread and tangy farmhouse cheese and curries, but also a family favourite with fish and chips.

900g (2lb) cucumber
2 onions, peeled
1 green pepper
3 tablespoons cooking salt
425ml (15fl oz) cider vinegar
225g (8oz) granulated sugar
2 tablespoons whole mustard seeds
1 teaspoon ground ginger

Slice all the vegetables into thin slices and place in a large china or glass bowl. Sprinkle with the salt and cover with a clean cloth. Leave overnight.

Next day, drain the vegetables into a colander and rinse well under cold running water. Dry well in a clean cloth. Put the vinegar, sugar and spices in a large pan and heat slowly, stirring frequently until all the sugar has dissolved. Bring to the boil and then add the vegetables. Simmer gently for 5 minutes only, so that the vegetables keep their bright green colour, then spoon the pickle into hot, sterlised jars. Pour over the remaining vinegar to cover.

Seal tightly and store until required.

Preserved Lemons

Lemons have been preserved in salt for centuries and are an essential ingredient in fashionable Moroccan and North African dishes, such as lamb tagine. It is the peel that is the important part of preserved lemons: the pulp can either be used or discarded. Add to tagines, casseroles, couscous and salads. Limes can be preserved in the same way.

10 medium lemons
About 3 tablespoons coarse sea salt
12 black peppercorns
12 coriander seeds
6 whole cloves
1 large cinnamon stick, broken into
    4 pieces
4 bay leaves

Wash the lemons, then cut five of them into eight wedges. Remove all the pips.

Squeeze the remaining lemons, reserving the juice.

Firmly press a layer of lemon wedges into the bottom of a sterilised 600ml (1 pint) jar. Cover with 2 teaspoons of salt, a few peppercorns and coriander seeds, 1 or 2 cloves, a piece of cinnamon stick and 1 bay leaf. Press down another layer of lemon wedges and repeat the salting and spicing process. Continue until there is only about 2.5cm (1in) of space at the top of the jar.

Now pour in the lemon juice, seal the jar and leave to mature in a cool, dark place for one month, after which they will be ready to eat. Try to put the lemons somewhere you can keep an eye on them and turn the jar occasionally. They should keep for about one year, even when opened, as long as you don't contaminate the contents, so keep fingers out.

When you want to use the preserved lemons, dig them out with a very clean wooden implement, and give them a rinse in cold water to remove excess saltiness (and any harmless white, lacy mould), then cut them into pieces, discarding the pulp if you prefer.

Runner Bean Pickle

I came across this pickle when I moved to Cornwall, where it is very traditional. It is excellent and a marvellous way of dealing with a glut of runner beans, which seems to happen to me every year. Don't be tempted to use stringy beans: they must be young and tender.

1½ tablespoons dry mustard powder
1½ tablespoons ground turmeric
1½ tablespoons cornflour
850ml (1½ pints) malt vinegar
900g (2lb) runner beans, trimmed and sliced
4 medium onions, peeled and finely sliced
4 garlic cloves, peeled and sliced
1 tablespoon cooking salt
700g (1½lb) demerara sugar
6 dried red chillies

Mix the mustard, turmeric and cornflour with a little of the vinegar to make a smooth paste. Place the beans, onions, garlic and salt in a pan and add the remaining vinegar. Simmer until the beans are tender. Stir in the spicy paste, sugar and chillies and heat gently, stirring frequently until the sugar has dissolved. Bring to the boil, stirring, then simmer for 20–30 minutes, until thick. Spoon the pickle into warm, sterilised jars, then seal. Store for two to three months before eating.

Sweet Pickled Onions

This is the best recipe I have ever used for pickling onions. The resulting onions are sweet and crisp. Do be careful not to cut away too much when you trim the tops and roots of the onions or they will disintegrate and become soggy. Do not worry if any small yellow spots appear on your pickled onions, they are perfectly harmless.

1·35kg (3lb) small pickling onions, trimmed
50g (1¾oz) cooking salt
2–3 sprigs of fresh tarragon
1–2 fresh green or red chillies, halved
1·5 litres (2¾ pints) white wine vinegar
425g (15oz) granulated sugar

Put the onions into a large bowl and cover with boiling water. Leave for about 20 seconds and then pour off the water. Cover with cold water, then peel the onions, keeping them under the water when peeling. Put them into a bowl, sprinkling each layer with salt. Cover with a clean cloth and leave overnight.

Rinse well and shake off as much water as possible. Pack into sterilised jars, adding a washed sprig of tarragon and half a chilli to each jar.

Boil the vinegar and sugar together for 1 minute. Pour the hot vinegar over the onions. Seal and store for two to three weeks before eating. Use within six months.

Spiced Quinces

These are wonderful with roast pork, duck and goose, or pork chops, ham, pâtes and terrines made from pork, duck or goose – and with cheese. It is difficult to give exact quantities in the recipe ingredients because they depend on the size of your quinces. I have used eight quinces as a guide, but use as many as you like, as the pickle will be very popular.

8 quinces
Cold water to cover
1 heaped dessertspoon sea salt
Coriander seeds
Cumin seeds
Granulated sugar
White wine or cider vinegar

Wash the quinces and rub off any fluff on their skins, then peel and core them. Cut each quince into eight pieces or so, depending on size, then place in a pan. Cover with water and add the salt, then bring to the boil. Simmer for about 10 minutes, then strain the liquid from the fruit, reserving it. Return the fruit to the pan.

Measure the cooking liquid in a measuring jug and for every 600ml (1 pint), add 450g (1lb) sugar, 150ml (¼ pint) vinegar, 1 teaspoon coriander seeds and ½ teaspoon cumin seeds, both of which have been gently roasted in a heavy frying pan.

Pour this liquor over the quinces and bring to the boil. Simmer for about 10 minutes, or until the fruit is tender. Using a slotted spoon, transfer the quince pieces to warm, sterlised jars. Continue cooking the liquor for a further 10 minutes, until you have a thick syrup. Pour this over the quince pieces and seal the jars.

The pickle will be ready to eat in two to three weeks if you can wait that long!

Sweet Pickled Peaches

This recipe dates back to the days when it was fashionable to grow your own peaches, apricots and nectarines. The fruits were very precious and had to be preserved for the winter. Serve with cold or hot ham, pork, duck or smoked poultry.

900g (2lb) firm peaches
Juice of ½ lemon
1 tablespoon whole cloves
300ml (½ pint) Sweet Pickling Vinegar
    (see page 62) made with white wine
    or cider vinegar

Place the peaches in a bowl and pour boiling water over them. Leave for about 1 minute, then drain. Peel the fruit – the skins will slip off easily. Brush the peaches with lemon juice to prevent them from going brown. Halve and stone, brushing with lemon juice. Crack a few stones with a nutcracker, remove and reserve the kernels. Stud each peach half with 2–3 cloves.

Put the vinegar in a large pan with the reserved kernels and bring to the boil. Boil for about 5 minutes, then carefully add half the peaches. Simmer very gently until they are just tender, but do not let them disintegrate. Remove the peaches with a draining spoon and place, cut sides down, in warm, sterilised jars. Poach the remaining peaches in the same way and place in the jars.

Boil the syrup for about 5 minutes until it has thickened and reduced, then pour it over the peaches. Seal and keep for one month before eating.

VARIATION
Nectarines or apricots can be pickled in the same way, but there is no need to skin apricots.

Spiced Orange Slices

Oranges were preserved in this way in the days when citrus fruit was scarce and expensive. These are delicious served with ham, pork or game.

10 large thin-skinned oranges, cut into 5mm (¼in) slices
600ml (1 pint) white wine vinegar
1kg (2¼lb) granulated sugar
1¼ cinnamon sticks
8g (¼oz) whole cloves
6 blades of mace

Place the orange slices in a large pan and cover with cold water. Simmer gently, partially covered with a lid, until the peel is tender, about 40 minutes (the peel should be very soft when pressed between finger and thumb).

Meanwhile, put all the other ingredients in a pan and boil for a few minutes to make a syrup. Drain the oranges, reserving the liquor, and place in the prepared syrup. Simmer, uncovered, for another 30–40 minutes, until the oranges are translucent. Transfer to a bowl and leave overnight.

Next day, boil the oranges in the syrup until they are thoroughly cooked. Pack the oranges into warm, sterilised jars, adding syrup to cover. Seal and leave for six to eight weeks before eating. The flavour is even better if left for several more months. Use the reserved liquor for topping up during storage as the oranges absorb the syrup.

Spiced Prunes

This preserve can be made at any time of the year. Try to use large succulent prunes; they are very good with cold pork, ham and hot or cold goose or game.

450g (1lb) large prunes
1·2 litres (2 pints) cold, fresh tea
450ml (¼ pint) red wine vinegar
225g (8oz) granulated sugar
1 teaspoon pickling spice, tied in a
    piece of muslin
Pared rind of 1 lemon

Soak the prunes overnight in the cold tea. Transfer them to a pan with half the soaking liquor and bring to the boil. Cover and simmer very gently for 10–15 minutes or until the prunes are just tender. Drain the prunes, reserving the liquid, and set aside until cold.

Put the vinegar, sugar, muslin bag of spices and lemon rind in a pan and heat gently, stirring frequently until the sugar has dissolved. Bring to the boil and boil rapidly for 5 minutes. Add 300ml (½ pint) of the reserved cooking liquid from the prunes and bring back to the boil.

Pack the cold prunes into warm, sterilised jars and pour the hot vinegar over them. Seal immediately. These prunes can be eaten 24 hours after pickling.

Ginger Pickled Pears

When selecting the pears, choose small pretty fruit with stalks. Deep golden in colour and sharp-sweet in flavour, these are super with goose or duck and a must with the Christmas ham. I have also served them with clotted cream for pudding.

700g (1½lb) hard Conference pears
225g (8oz) light muscovado sugar
175ml (6fl oz) red wine vinegar
1 thin-skinned lemon
2·5cm (1in) piece of fresh root ginger,
    peeled and finely sliced
2·5cm (1in) cinnamon stick
2 small star anise

Preheat the oven to 160°C, 325°F, gas mark 3. Peel and halve the pears, leaving the stalks on. Scoop out the cores with a melon-baller or teaspoon, then place in a casserole dish.

Put the sugar and vinegar in a pan. Slice the lemon thinly and add to the pan with the spices. Stir over a gentle heat to dissolve the sugar, then bring to a simmer. Pour over the pears, then cover the casserole dish with a lid and bake for about 40 minutes, until the pears are tender when tested with a skewer.

Leave to cool in the casserole for 24 hours, then spoon into sterilised jars, adding syrup to cover, or store in plastic lidded containers (they will keep for two to three months in the fridge). The pears need at least 48 hours to take colour and flavour, and are best kept for at least one week before using.

Pickled Mushrooms

450g (1lb) small button mushrooms
300ml (½ pint) white wine vinegar
1 shallot, peeled and sliced
1 tablespoon finely chopped fresh
    root ginger
2 blades of mace
1 teaspoon table salt
½ teaspoon ground black pepper
½ teaspoon freshly grated nutmeg
4 sprigs of fresh thyme
4 tablespoons sherry

Wipe the mushrooms and trim the stalks (use for flavouring soups or stews). Pour the vinegar into a pan and add the remaining ingredients, except for the sherry. Bring to the boil, then simmer for 5 minutes. Add the mushrooms to the boiling vinegar, and bring back to the boil. Cover the pan and simmer for 1 minute until the mushrooms have shrunk slightly. Stir in the sherry, then use a draining spoon to pack the mushrooms into warm, sterilised jars and pour in the hot vinegar. Seal at once while still hot, then leave for one to two weeks before eating. Use up quickly.

Sweet Vegetable Pickle

A good pickle to make in the early autumn when there is a glut of courgettes. It is excellent with cold meats, pâtes, terrines and cheese.

700g (1½lb) firm courgettes
2 large onions
1 celery heart
1 large cauliflower
225g (8oz) cooking salt
1.2 litres (2 pints) warm water
600ml (1 pint) cold water
450g (1lb) Bramley or other cooking apples
25g (1oz) garlic, peeled and crushed
1.2 litres (2 pints) white vinegar
50g (1¾oz) cornflour
700g (1½lb) dark muscovado sugar
1 tablespoon ground cinnamon
1 tablespoon ground cumin
1 tablespoon ground turmeric
1 teaspoon ground allspice
1 teaspoon freshly grated nutmeg

Wash and prepare the vegetables. Cut the courgettes, onions and celery into 1cm (½in) dice and place in a large glass or china bowl. Cut the florets from the stem of the cauliflower and add to the bowl.

Mix the salt with the warm water, stirring until it dissolves. Add the cold water and pour over the prepared vegetables. Leave to soak overnight, then drain and rinse well under cold running water. Drain on a clean cloth while you continue with the recipe.

Peel, core and dice the apples. Put the apples and garlic in a large pan with 300ml (½ pint) of the vinegar. Cook until soft.

Mix the cornflour with about 4 tablespoons of the remaining vinegar to make a thin paste and set aside. Add the rest of the vinegar, the sugar and the spices to the apple mixture. Slowly bring to the boil, stirring until the sugar has dissolved. Mix a few spoonfuls of the hot liquid into the cornflour paste and then stir everything back into the pan. Stir continuously while the mixture thickens to give a smooth sauce. Simmer for 2–3 minutes, then add the drained pickled vegetables. Cook for 5 minutes, then spoon into warm, sterilised jars. Seal and store for at least two weeks before using. Use within two to three months.

Pickled Eggs

These became common in Britain at the beginning of the eighteenth century and are still a staple in fish and chip shops and some pubs. Eat with cold meat or cheese, although they are even better as a snack, with a glass of real ale, or as an appetiser, dipped into celery salt.

Pullets' eggs make a delicious mouthful or two; and quails' eggs are fun; but whatever eggs you choose, make sure they are not newly laid, as they will not peel properly. Adjust the boiling time for different sized eggs.

Eggs
Spiced Vinegar (see page 60)
Small sprigs of fresh thyme (optional)

Place the eggs in a pan of cold water (enough to just cover them). Bring to the boil, stirring the water gently to help keep the yolks in the centres, then boil for about 8 minutes, depending on the size of the eggs. Drain and plunge immediately into cold water to stop the yolks turning black.

When cool, shell the eggs, removing the skin with the shell. Pack into sterilised, wide-mouthed jars, then cover with cold Spiced Vinegar. Add a few sprigs of fresh thyme, if you wish, then seal and keep for at least one week before using.

Pickled eggs improve with keeping; store for up to one year in the fridge or in a cool larder.

Pickled Green Walnuts

The walnut tree was introduced into Britain by the Romans, but sadly there has been a steady decline in the number of specimens since the eighteenth century. If you are lucky enough to know of a source of young walnuts, you can make this delectable pickle. It was at its height of popularity in Victorian London, where it was a favourite accompaniment to grilled mutton chops, cold roast beef in sandwiches or with Stilton cheese. Pickled walnuts were also put into casseroles and steak and kidney puddings. Try them like this – you will not be disappointed.

Pick your walnuts in June or July, when they are still green and before the shells have begun to form. Remember to wear rubber gloves while you are preparing them, unless you want a pair of beautiful mahogany hands!

450g (1lb) freshly picked green walnuts
1·2 litres (2 pints) water
100g (3½oz) cooking salt
600ml (1 pint) Spiced Vinegar (see page 60)

Prick the walnuts lightly with a large needle so that the pickle will permeate them. Discard any nuts with hard patches at the end opposite to the stalk, the point where the shell starts to form.

Make a strong brine by mixing half the water and half the salt together in a bowl. Add the walnuts and leave to stand for three to four days. Drain the nuts. Mix up another brine using the remaining water and salt. Put the walnuts into this and leave for a further week.

Drain the nuts well, rinse and dry them, then spread them out on a tray or plate. Leave the walnuts in a sunny place for two days or until they are completely black, then pack them into sterilised jars. Heat the Spiced Vinegar and pour it over the walnuts to cover them completely. When cold, seal tightly and store in a cool, dry place for six to eight weeks before using.

VARIATION
Sweet Pickled Walnuts
Prepare the nuts as above, but cover with Sweet Pickling Vinegar (see page 62).

Flavoured Sugars, Salts and Syrups

Of all the items on the medieval spice account, sugar was the one destined to have the greatest effect on Britain's eating habits. The Romans had known it only as a medicine, but now it came into its own in the kitchen. As well as plain sugars of several grades, rose- and violet-flavoured sugars were imported, and those who could afford them consumed large quantities. Apparently, in 1287 the royal household used 677lb of sugar, 300lb of violet sugar and 1,900lb of rose sugar.

Flavoured sugars are a delightful addition to the modern store cupboard and are so simple to make. Use them to flavour plain puddings, custards, milk puddings, biscuits and cakes. It is well worth experimenting with different herbs, flowers and flavourings.

Recipe books from the 18th and 19th centuries often include recipes for concentrated syrups, which were intended to be diluted with water when required, rather like modern squashes. Many were the descendants of the apothecary's cordials, extolled for their virtues in Elizabethan books of remedies, and became favourite drinks for ladies at balls and dinners. They are an excellent way of sweetening and flavouring desserts such as ice cream, sorbet, yogurt, fools, trifles and custards. Flavoured syrups can also be used hot or cold as a sauce for a steamed pudding or ice cream, or diluted with water or milk as a drink.

Fruit Syrups

These are excellent for diluting to make refreshing summer drinks or 'medicines' for a sore throat in the winter.

Blackcurrant Syrup

The best fruits for making syrups are blackcurrants, raspberries, blackberries, loganberries, strawberries and elderberries. Redcurrants make rather an astringent syrup, so mix the fruit with strawberries or raspberries. If you are making berry syrup, use half the amount of water that is used in this recipe.

Dilute to make jellies and drinks, or use as a sauce for ice creams, milk puddings and steamed or baked sponge pudding.

900g (2lb) ripe blackcurrants
300ml (½ pint) cold water
Granulated sugar

Discard any leaves, stalks and damaged fruit, then wash. Put the blackcurrants in a large china bowl over a pan of simmering water and add the water. Cook for about 1 hour, pressing the fruit frequently with a wooden spoon to release the juice. Strain through a jelly bag overnight (see page 129).

Measure the strained juice and pour into a pan. For every 600ml (1 pint) juice, add 450g (1lb) sugar. Stir over a low heat until the sugar has completely dissolved and then bring to the boil. Simmer for 15 minutes.

Pour the hot syrup into small, sterilised, screw-cap bottles, filling them to within 2·5cm (1in) of the top, then screw the tops on loosely. Sterilise in the usual way (see page 11).

Tighten the screw caps and leave to cool. When the syrup is cold, label and store in a cool, dark, airy cupboard. It will keep well for up to two months.

Mulled Wine Syrup

This spicy syrup is a great base to have in the fridge at Christmas to save complicated last-minute preparations, especially when unexpected guests arrive. Add to your red wine as it heats up, to suit your personal taste.

6 large oranges
225g (8oz) granulated sugar
1 teaspoon ground cinnamon
1 teaspoon freshly grated nutmeg
1 dessertspoon whole cloves
1·75 litres (3 pints) cold water

Roughly cut up the oranges and place in a pan with the sugar, spices and water. Bring slowly to the boil, stirring frequently, until all the sugar has dissolved and then simmer gently, uncovered, for about 30 minutes. Strain through a muslin-lined sieve and then pour into a screw-cap sterlised bottle. Leave until completely cold and then store in the fridge until required.

Rosehip Syrup

Conserves and syrups of rose hips were once made in country house still-rooms. An economical eighteenth-century recipe in the Dryden family papers at Canons Ashby, near Northampton, makes a conserve from the pulp of 'half a peck of Hipps' and then uses the leftover skin and pips to make syrup.

Try rosehip syrup diluted with hot water as a soothing drink when you have a cold, or drizzle over meringues, milk puddings, fruit fools, creams, blancmanges and fruit salads. Pick the largest rose hips you can – the fruit of the *Rosa rugosa* is ideal – as they are juicier.

700g (1½lb) large ripe rose hips
2 litres (3½ pints) cold water
350g (12oz) granulated sugar

Mince the rose hips or process coarsely. Place in a pan with 1.2 litres (2 pints) of the water. Bring to the boil, then remove from the heat. Cover the pan and leave to infuse for 15 minutes. Strain through a jelly bag (see page 129), then return the pulp to the pan, adding the remaining water. Bring to the boil, then cover and put aside for another 15 minutes. Strain this juice through the jelly bag as before. Put into a clean pan, bring to the boil, and then boil until reduced to about 600–750ml (1–1¼ pints). Add the sugar, stir until dissolved, then boil for a further 5 minutes. Bottle and sterilise as for Blackcurrant Syrup (see page 88).

Flavoured Sugars

Orange or Lemon Sugar

This is good for flavouring cakes, biscuits, puddings and creamy desserts, as it saves grating orange or lemon rinds.

Set the oven to its lowest temperature. Remove the rind from the fruit with a potato peeler without removing any pith. You will need about 100g (3½oz). Spread out on a baking sheet covered with kitchen foil. Place in the oven and leave for about 3 hours, or until dried out. Allow to cool. Layer the sugar with the dried rind in a clean, airtight jar. Seal and label, then give the jar a shake. Keep in a cool, dark place for at least one week before using to allow the flavour to develop.

Vanilla Sugar

Vanilla pods are the fruit of a climbing orchid. Their flavour develops during the curing process. The pods can be used to flavour the foods as they are, or the flavour can be extracted from them to give vanilla extract or essence. When you are buying a vanilla pod, try to find one which is covered with crystals of vanilla as this indicates freshness.

Use vanilla sugar to flavour creams, custards, cakes, whipped cream, chocolate dishes and to make French Cherry Jam (see page 111) and Raspberry and Vanilla Jam (see page 122).

1 fresh vanilla pod
450g (1lb) caster sugar

Split the vanilla pod to expose the seeds for full flavour, then place in a lidded jar with the caster sugar. Shake, then leave for one week. Shake again and leave for two to three weeks before using. Top up with more sugar when necessary.

Spiced Sugars

Many spices can be used to flavour sugar. Spiced sugars are lovely used in fruit pies and puddings, cakes and biscuits instead of plain sugar. Try cinnamon sticks, caraway seeds, whole cloves, allspice berries or dried root ginger.

Cinnamon Sugar

Use to sprinkle on cakes, biscuits and puddings, or cream with butter and spread on hot fresh toast to make cinnamon toast.

450g (1lb) caster sugar
50g (1¾oz) whole cinnamon sticks

Tip the sugar into a clean airtight jar, then press the cinnamon sticks down into the sugar. Seal and label, then give the jar a good shake. Keep in a cool, dark place for at least one week before using to allow the flavour to develop.

Herb Sugars

Herbs recommended for perfuming sugar include angelica, bay, blackcurrant leaves, rose geranium, and other scented leaf geraniums, hyssop, lavender, lemon balm, lemon verbena, mint (especially eau-de-cologne and pineapple), rose petals, rosemary, elderflowers, sweet cicely and sweet violet.

Lavender Sugar

This is delicious in ice cream or any milk or creamy pudding and makes wonderfully scented scones, biscuits and cakes. The sugar will be very fragrant so use with caution, replacing only a small amount of the sugar required in a recipe with Lavender Sugar, or just sprinkle on top of biscuits, pies and cakes after cooking. Also use with butter to glaze carrots. Pick the lavender when it is just coming into flower.

225g (8oz) caster sugar
50g (1¾oz) sprigs of fresh lavender flowers (insect free) or 25g (1oz) dried lavender flowers

Layer the sugar with the lavender flowers in a lidded jar. Leave in a warm place for one to two weeks, giving the jar a shake now and again to distribute the scent evenly. When the sugar has absorbed the scent of the lavender, sift it to remove the lavender flowers. Keep the jar tightly sealed and use as required.

VARIATIONS
Rose Petal Sugar
Layer the sugar with 50g (1¾oz) petals of highly scented, unsprayed garden roses instead of lavender. Use to flavour creamy desserts, cakes, puddings and biscuits.

Mint Sugar
Use sprigs of mint instead of lavender. Makes a delicious ice cream.

Borage Sugar

Borage is very simple to grow from a packet of seeds. Scatter them over a patch of sunny, well-drained soil in spring or autumn and plants will emerge every year thereafter. The bright blue, star-like flowers make a pretty garnish to fruit salads and fruit cups, such as Pimms, and can be crystallised. They can also be used to flavour sugar for fruit salads and desserts.

100g (3½oz) freshly picked borage flowers
300g (10½oz) caster sugar

Wash the flowers and dry carefully. Grind them with the sugar in a mortar, or process, until completely mixed. Spread out on a baking tray and dry in the sun, an airing cupboard or in a very cool oven. When completely dry, store in an airtight jar and use as required.

Herb Salts

Sea salt can be flavoured with herbs to add to savoury dishes for extra piquancy. Experiment with bay, rosemary, tarragon, thyme, fennel, sage, basil and marjoram.

Tarragon Salt

50g (1¾oz) sprigs of fresh tarragon
100g (3½oz) sea salt

Set the oven to its lowest temperature. Strip the tarragon leaves from the stalks and discard the stalks. Coarsely chop the leaves and place in a blender. Add the sea salt and process until the leaves are finely chopped. Cover a baking sheet with kitchen foil and spread the herb mixture out on it. Place in the oven, leaving the door ajar. Leave for about 1½ hours until crisp and dry. Allow to cool, and then pack into a clean airtight jar. Cover and label. Use as required.

Bay Salt

1–2 dried bay leaves
100g (3½oz) sea salt

Put a layer of salt into a clean airtight jar. Add one or two bay leaves, depending on size, then top up with the remaining salt. Cover and label, then leave to infuse for several weeks before using.

Peppercorn and Rosemary Salt

15g (½oz) sprigs of fresh rosemary
½–¾ tablespoon mixed peppercorns
100g (3½oz) sea salt

Set the oven to its lowest temperature. Strip the rosemary leaves from the stalks and discard the stalks. Mix with the peppercorns and salt in a blender and process until the rosemary is finely chopped. Continue as for Tarragon Salt (see page 102).

VARIATION
Peppercorn and Thyme Salt
Use sprigs of thyme, sage, oregano or marjoram instead of rosemary. Continue as above.

Herb and Flower Syrups

The best herbs and flowers to use for making a flavoured syrup are those with a fairly strong scent: angelica, elderflower, fennel, rose geranium and scented-leaf geraniums, hyssop, lavender, lemon balm, lemon verbena, all mints, pineapple sage, rose petals, rosemary, sweet cicely and sweet violet.

Pineapple Sage Syrup

This very attractive herb makes a delicious syrup for pouring over homemade pineapple ice cream or syllabub, or over upside-down pineapple pudding. It is less hardy than the normal sage, so will need protecting from frosts, but its strong pineapple-scented leaves and lovely red flowers make it well worth growing.

350g (12oz) granulated sugar
300ml (½ pint) cold water
4–6 sprigs of pineapple sage

Dissolve the sugar in the water over a very gentle heat, stirring frequently, and then bring to the boil. Add the pineapple sage, pushing it down into the syrup, and bring back to the boil. Remove from the heat. Cover with a clean cloth and leave until completely cold.

Strain through a muslin-lined sieve and pour into a sterilised screw-cap bottle. Keep in the fridge and use as required.

VARIATION
Herb and Flower-scented Wine Syrups
Instead of using all water, use half medium dry white wine and half water. Make as above.

Elderflower Syrup

This concentrated elixir of the Muscat-scented elderflower blossom is probably the easiest and most useful way to capture for use during the months when it is out of season. Use as the basis for elderflower cordial (below), to flavour batters, milk puddings, custards, creams, fools, ice creams and fruit salads, and to flavour sauces to accompany baked and steamed sponge puddings.

700g (1½lb) granulated sugar
Grated rind and juice of 1 lemon
600ml (1 pint) cold water
12–18 freshly picked, opened
    elderflower heads

Put the sugar in a pan with the grated lemon rind and the water. Heat slowly, stirring frequently until all the sugar has dissolved, then bring to the boil.

Shake the elderflower heads clean and add the blossoms to the pan, cutting off the stalks close to the creamy heads. Push the flowers down into the syrup and bring back to the boil. Remove from the heat, cover with a clean cloth and leave until completely cold.

Stir the lemon juice into the syrup. Strain through a damp muslin-lined sieve, twice, and then pour into sterilised screw-cap bottles. Store in the fridge and use as required.

To Make Elderflower Cordial

Mix shortly before drinking. Simply pour a slug of elderflower syrup into a chilled glass. Stir in a little freshly squeezed lemon juice and dilute to taste with chilled water. Garnish with lemon slices and ice cubes. (The ratio of lemon to syrup is a matter of personal taste, but try one part lemon to three parts syrup to start).

To make Elderflower Cordial from fresh elderflowers, see the recipe on page 220.

Jams and Jellies

Jellies have been made from apples and soft fruit, such as raspberries, strawberries, mulberries and barberries, since Tudor and Stuart times. They were strained through a linen cloth or jelly bag, and boiled until they set with their own pectin content. They were regarded as accompaniments to meat, game and poultry rather than a spread for bread.

The word 'jam' did not reach the printed cookery books until 1718. Thereafter both the name and the method of preparation became common. By the late nineteenth century jam was being manufactured outside the home, using the surplus of fruit and vegetables during the agricultural depression. Jam, spread on white bread and margarine, became part of the poor man's staple diet.

Some of the National Trust's restaurants make their own jams and jellies, using estate or locally grown fruit, to serve with their famous cream teas. I have included some of these recipes.

Jam and jelly making is hugely enjoyable and very satisfying, but it does become easier with experience. The tendency at first is to over-boil and caramelise the sugar, which does not quite spoil the jam, but does ruin its colour. You have to watch it very carefully as it boils and be sure to take the pan off the heat while you test for a set. Do not be put off if your first jar of jelly or jam is not perfect. Make in small quantities.

Jams

Jam is a mixture of fruit and sugar cooked together until set. Pectin, a gum-like substance that occurs in varying amounts in the cell walls of fruit, is essential to setting. Acid is also necessary to help release the pectin, improve the colour and flavour, and prevent crystallization. Fruits rich in pectin and acid are blackcurrants, cooking apples, crab apples, cranberries, damsons, gooseberries, lemons, limes, Seville oranges, quinces and redcurrants. Fruits containing a medium amount are dessert apples, apricots, bilberries, blackberries, greengages, loganberries, mulberries, plums and raspberries. Fruits low in pectin, which do not give a good set unless mixed with other high-pectin fruit, are cherries, elderberries, figs, grapes, japonica, medlars, nectarines, peaches, pears, rhubarb, rowanberries and strawberries. If acid is the only ingredient required and the fruit has enough pectin, the most convenient way of adding it is in the form of lemon, redcurrant or gooseberry juice.

To Make Jam

Choosing and preparing the fruit
- It should be dry, fresh and barely ripe. Over-ripe fruit does not set well as it is low in pectin.
- Pick over the fruit, discarding any damaged parts, and wash or wipe it.
- Simmer gently until soft and reduced by about one-third to break down the cell walls and release the pectin. Make sure that all fruit skins are completely soft, especially thick-skinned fruits such as blackcurrants, before adding the sugar, because this will instantly toughen them.

Choosing and using the sugar
- There is no keeping difference between jams made with beet sugar and sugar cane.
- Any kind of sugar (except icing sugar) will make jam, but preserving sugar makes a slightly clearer jam which needs less stirring, and produces far less scum. Brown sugars can mask the fruit flavour.
- Commercial jam sugar, based on granulated sugar with added pectin and acid, guarantees a set for any fruit. Choose the type that guarantees a set in 4 minutes with no testing. This is useful for low-pectin fruits but is expensive.
- Warm the sugar in the oven before adding to the fruit to reduce the cooking time. Take the pan off the heat before adding the sugar and stir until completely dissolved. If the mixture boils before the sugar is dissolved, it will crystallise and the jam will be crunchy and spoilt. Once the sugar is dissolved completely, bring the jam to a rolling boil (the boiling continues when the jam is stirred with a wooden spoon).

Testing for a set
Most jams reach setting point after 5–20 minutes' boiling, but always start testing for a set after 5 minutes as over-boiling spoils the colour and flavour.

- Sauce Test: Before you start making your jam, put a saucer into the fridge or freezer to get cold. Remove the jam from the heat and put about 2 teaspoons onto the cold saucer. Allow it to cool, then push your fingertip across the centre of the jam. If the surface wrinkles well and the two halves remain separate, setting point has been reached. If not, return the pan to the heat and boil again for 5 minutes, then test again.
- Temperature Test: If you have a sugar thermometer, hold it in the boiling jam, without resting it on the bottom of the pan. Bend over until your eyes are level with the 104°C (220°F) mark on the thermometer: this is the temperature the jam should reach when it is at setting point.
- Flake Test: To check setting point, do a quick 'flake' test. Dip the bowl of a cold wooden spoon in the jam. Take out and cool slightly, then let the jam drop from the edge of the spoon. At setting point, the jam runs together, forming flakes, which break off cleanly with a shake of the spoon.

Potting and covering
1. Wash the jars in very hot water and dry in the oven at 140°C, 275°F, gas mark 1. Leave them there until you are ready to pot the jam. The jars must be warmed before filling, or the hot jam will crack them.
2. Once the setting point has been reached, pot the jam immediately, except for strawberry and raspberry jam and all marmalades. Leave these to stand for 10–15 minutes to let the fruit or rind settle, to prevent it from rising to the surface in the jar.
3. Pour the jam into the jars using a ladle or small jug, filling them almost to the top, leaving no space for bacteria to grow. A jam funnel makes filling easier.
4. Cover the jam immediately with waxed paper discs, placing the waxed sides down on the surface of the preserve. The surface of the jam should be completely covered by the waxed paper (buy the right size for the type of jar used). Press gently to exclude all air. Then immediately add lids or dampened cellophane covers (damp side downwards) and secure with rubber bands.
5. For every 1.35kg (3lb) sugar, the yield will be about 2.25kg (5lb) jam.

Storing
1. Stand the jars of jam aside until completely cold, then label clearly with type of jam and the date it was made.
2. Store in a cool, dark, dry and airy cupboard. Homemade jam and jelly will keep well for up to one year, but may deteriorate in colour and flavour if kept longer.

Strawberry Jam

This is the jam everyone loves, young and old, and is the traditional one served with National Trust cream teas. It is only worth making if you have a good supply of fresh fruit from your own garden, or direct from a pick-your-own farm. This recipe will take you several days, although it is very simple.

Strawberries are rather lacking in acid and pectin, so choose small, firm berries, preferably of the more acid varieties, and use jam sugar to ensure a good set. If you have any problems getting the jam to set, add the juice of 1 lemon when bringing the fruit and sugar to the boil and continue as recipe.

1·5kg (3lb 5oz) small strawberries
1·5kg (3lb 5oz) jam sugar

Hull the strawberries, but do not wash them, and then layer with two-thirds of the sugar in a wide, shallow china or glass bowl. Sprinkle over the remaining sugar. Cover with a clean cloth and leave at room temperature (as long as your room is not too warm) for 24 hours.

Next day, scrape the contents of the bowl into a pan and bring slowly to the boil. Allow the mixture to bubble over a low heat for 5 minutes and then remove from the heat, cover with a clean cloth and leave for a further 48 hours.

Return the pan to the heat and bring back to the boil. Boil for about 10–15 minutes until setting point is reached (see page 109). Remove the pan from the heat and skim off any scum. Allow the jam to cool for 10 minutes and then give it a good stir, so that the fruit is well-dispersed – it will then remain suspended rather than rising to the top of the jam. Pot and cover in the usual way.

Black Cherry Conserve

Cherries are low in pectin and acid, so you need to add lemon juice, although I like a rather soft set for this jam. If you are lucky enough to have Morello cherries, which are more acid, use just ½ lemon.

1kg (2¼lb) cherries
750g (1lb 10oz) granulated sugar
1 lemon

Wash and stone the cherries – you should have around 900g (2lb) when prepared – and then layer with the sugar in a bowl. Cover with a clean cloth and leave for 24 hours.

Next day, peel the lemon and squeeze out the juice. Tie the lemon peel and pith loosely in a square of muslin. Place the cherry and sugar mixture in a pan with the lemon juice and the bag of peel. Bring slowly to the boil, stirring frequently, then boil rapidly until setting point is reached (see page 109). Remove and squeeze the muslin bag and then discard. Skim the jam and then leave to cool for about 10 minutes, then pot and cover in the usual way.

Variations
French Cherry Jam
Use Vanilla Sugar (see page 96) and crush the cherry stones before putting in the muslin bag with the lemon peel. Continue as above.

Black Cherry and Walnut Conserve
Stir in 50g (1¾oz) chopped walnuts when setting point is reached. Allow to cool and then pot.

Cliveden Rhubarb and Ginger Jam

1·8kg (4lb) rhubarb, trimmed and cut into 2.5cm (1in) pieces
1·8kg (4lb) granulated sugar
Grated rind and juice of 2 lemons
50g (1¾oz) dried root ginger, bruised
50g (1¾oz) preserved stem ginger, chopped

Put the rhubarb in a large bowl, layering it with the sugar, the lemon rind and juice. Cover with a clean cloth and leave to stand overnight.

Next day, turn the soaked fruit and all its juice into a preserving pan and add the bruised root ginger tied in a piece of muslin. Simmer gently until the fruit is soft and pulpy, stirring frequently to dissolve the sugar completely. Boil rapidly to setting point (see page 109). Remove the muslin bag of ginger. Stir in the chopped stem ginger, then pot and cover in the usual way.

Rhubarb and Rose-petal Jam

Use freshly picked, unsprayed, scented rose petals, red if possible, from the garden. Do not buy roses from a florist, as they will have been sprayed. Red roses give a better flavour and a lovely colour.

450g (1lb) rhubarb
Juice of 1 lemon
450g (1lb) granulated sugar
2 handfuls of scented rose petals
    (about 5 roses)

Wipe the rhubarb and cut into 1cm (½ inch) pieces. Put in a shallow china dish, add the lemon juice and then cover the rhubarb with the sugar. Cover with a clean cloth and leave to stand overnight.

Next day, remove the white bit, or heel, from each rose petal and discard. Cut the rose petals into strips and put in a pan with the rhubarb mixture. Bring to the boil slowly, stirring until the sugar has dissolved. Boil briskly until setting point is reached (see page 109).

Leave to cool slightly and then pot and cover in the usual way.

VARIATION
Rhubarb and Elderflower Jam
Tie 3 elderflower heads in a piece of muslin and add to the rhubarb, lemon juice and sugar overnight. Next day, add the grated rind of the lemon and boil until set as above.

Blackcurrant Jam

Blackcurrants make a gorgeously rich jam with a wonderful flavour, but make sure the initial cooking is thorough so that the skins are soft. Taste every now and again to check that the currants are well cooked before you add the sugar.

900g (2lb) blackcurrants
600ml (1 pint) water
1·35kg (3lb) granulated sugar

Wash the currants and remove the strings and stalks. Place in a pan with the water and bring slowly to the boil. Reduce the heat and simmer gently for about 20 minutes, or until the fruit is tender, stirring occasionally.

Add the warmed sugar, stir until completely dissolved and then boil rapidly for about 5 minutes or until setting point is reached (see page 109). Pot and cover in the usual way.

VARIATIONS
Blackcurrant and Apple Jam
Cook 350g (12oz) peeled, cored and sliced apples in 300ml (½ pint) water until tender. Add to the cooked blackcurrants and continue as above.

Blackcurrant and Rhubarb Jam
Cook 450g (1lb) rhubarb in 300ml (½ pint) water until pulpy. In another pan cook 450g (1lb) blackcurrants until tender. Mix the two fruits together and continue as above.

Blackberry and Apple Jam

This is one of a number of jams that the cooks at Cliveden, Buckinghamshire, make for their visitors. Use windfall or crab apples for a very economical jam.

900g (2lb) blackberries
150ml (¼ pint) water
450g (1lb) cooking apples
1·35kg (3lb) granulated sugar

Place the blackberries in a large pan and add half the water, then simmer for about 15 minutes until tender. Press through a fine nylon sieve to remove all the pips, if you wish, but be sure to press through all the fruit pulp and juice for the jam. Peel, core and roughly chop the apples. Simmer gently in a separate pan with the remaining water for about 10 minutes until soft and pulpy. Place in a large pan with the blackberry pulp and warmed sugar and stir over a low heat until the sugar has dissolved. Then bring to the boil and boil rapidly for 10–15 minutes until setting point is reached (see page 109). Pot and cover in the usual way.

VARIATION
Hedgepick Jam
This recipe is based on one provided by the Women's Institute during World War II, using fruits picked from the hedgerows. Any combination of fruits can be used but make sure that over half the quantity are fruits high in pectin, such as crab apples, blackberries and damsons.

Apricot Jam

This is my favourite jam, especially with freshly baked croissants. My sister Melanie makes it for me as a treat – dried apricot jam is just not the same. Apricots need extra acid to give a good set, so lemon juice is added.

1·35kg (3lb) fresh apricots
Juice of 1 lemon, strained
300ml (½ pint) water
1·35kg (3lb) granulated sugar

Halve and stone the apricots, reserving 12 stones. Using a nutcracker, crack open the reserved stones and remove the kernels. Blanch them in boiling water for 1 minute, then drain and transfer to a bowl of cold water. Drain again, then rub off the skins with your fingers.

Simmer the apricots and kernels gently in the lemon juice and water for about 15 minutes or until the fruit is soft and the water has reduced. Add the warmed sugar and stir until it has completely dissolved, then boil rapidly for 10–15 minutes, or until setting point is reached (see page 109). Pot and cover in the usual way.

Blueberry Conserve

Blueberries are distant relatives of the European bilberry and very easy to grow in your own garden.

600g (1lb 5oz) blueberries
850g (1¼lb) granulated sugar
Juice of 1 lemon
Tiny pinch of table salt

Put the blueberries into a china or glass bowl with half the sugar, all of the lemon juice and the salt. Stir well, then cover with a clean cloth and leave to stand overnight.

Next day, pour the contents of the bowl into a large pan with the remaining sugar. Stir over a low heat until the sugar has completely dissolved, then increase the heat and boil rapidly for 10–12 minutes until setting point is reached (see page 109). Skim off any scum, and then pot and cover in the usual way.

VARIATION
Blueberry and Lavender Jam
Make as above, but use Lavender Sugar (see page 100), or add 1 teaspoon individual lavender buds to the blueberries in the bowl.

Red Chilli Jam

My sister Marilyn invented this recipe after we had shared a basket of French fries with a delicious chilli-jam dip in a trendy wine bar in Plymouth. Her recipe really hits the spot. Use instead of mint jelly with lamb or serve with crab cakes and fried squid or with soft cheese, especially goat's cheese.

175g (6oz) onions, peeled and finely chopped
350g (12oz) sweet red peppers, deseeded and finely chopped
50g (1¾oz) medium red chillies, deseeded and finely chopped
2 large garlic cloves, peeled and chopped
450ml (¾ pint) white wine vinegar
350g (12oz) granulated sugar
2 tablespoons lemon juice

Cook the onions in a small quantity of boiling water for 5 minutes to soften them, then drain.

Put the peppers, chillies and garlic in a pan with the onions and vinegar and then simmer gently for about 25 minutes, until soft. Stir in the warmed sugar and continue to stir over gentle heat until it has completely dissolved. Add the lemon juice, stir and bring to the boil. Boil rapidly for about 15 minutes until setting point is reached (see page 109).

Pot and cover in the usual way. Red Chilli Jam will keep for about three weeks in the fridge and will get hotter the longer you keep it.

Raspberry Jam

You need slightly under-ripe raspberries to make really successful jam with no hint of mustiness or mould, so if you do not grow your own raspberries, the next best thing is to find a pick-your-own farm. Loganberries can be used instead.

1kg (2¼lb) raspberries
1kg (2¼lb) granulated sugar

Pick over the raspberries carefully, but do not wash them. Put them in a large, shallow china or glass dish and pour over the sugar. Cover with a clean cloth and leave for 24 hours, pounding them together every now and again.

Next day, tip the raspberry and sugar mixture into a pan and bring very slowly to the boil, stirring frequently. Boil fast for 3–5 minutes, or until setting point is reached (see page 109). Pot and cover in the usual way.

VARIATION
Raspberry and Vanilla Jam
Split 2 vanilla pods and add to the raspberry and sugar mixture, or use Vanilla Sugar (see page 96).

Uncooked Raspberry Conserve

1kg (2¼lb) raspberries
1·25kg (2lb 12oz) caster sugar

Preheat the oven to 180°C, 350°F, gas mark 4. Place the raspberries and sugar in separate ovenproof bowls. Put both bowls in the oven for about 20 minutes, until the contents of each is hot. Quickly and thoroughly stir the fruit and sugar together until all the sugar has dissolved, lightly crushing the fruit as you do so. Pot and cover in the usual way.

VARIATION
Uncooked Strawberry Conserve
Use strawberries instead of raspberries.

High Dumpsie Dearie Jam

Although an old Gloucestershire recipe, this jam was popular all over the country in Victorian times, when it was called Mock Apricot Jam.

900g (2lb) cooking apples, peeled and cored
900g (2lb) cooking pears, peeled and cored
900g (2lb) large plums
Juice of 1 large lemon, strained
300ml (½ pint) water
3 whole cloves
1 small cinnamon stick
2·75kg (6lb) granulated sugar

Cut the apples and pears into even-sized pieces. Halve and stone the plums, reserving the stones, then place all the fruit in a large pan with the lemon juice and water. Tie the reserved plum stones, cloves and cinnamon in a piece of muslin and add to the fruit. Simmer very gently until the fruit is soft. Remove and discard the muslin bag. Add the warmed sugar and stir until it has completely dissolved, then bring to the boil. Boil rapidly for about 15 minutes until setting point is reached (see page 109). Pot and cover in the usual way.

Cliveden Red Gooseberry and Elderflower Jam

A recipe used by the cooks at Cliveden's Conservatory Restaurant, just outside Maidenhead. Omit the elderflowers if you wish and substitute scented geranium leaves or other herbs.

1·35kg (3lb) gooseberries
300ml (½ pint) water
6–8 elderflower heads
1·35kg (3lb) granulated sugar

Place the gooseberries in a large pan and add the water. Simmer the fruit gently for about 20 minutes, or until pulpy. Meanwhile, snip the tiny flowers off the elderflower heads, making sure they are clean and insect-free. Stir the flowers and warmed sugar into the cooked fruit. Heat gently, stirring until the sugar has dissolved, and then boil rapidly for about 15 minutes until setting point is reached (see page 109). Pot and cover in the usual way.

VARIATIONS
Gooseberry and Orange Jam
Cook the gooseberries with the grated rind and juice of 3 oranges instead of the elderflowers.

Gooseberry and Redcurrant Jam
Use 900g (2lb) gooseberries and 450g (1lb) redcurrants and omit the elderflowers.

Gooseberry and Strawberry Jam
Use 700g (1½lb) gooseberries and 700g (1½lb) strawberries. Cook the fruits separately to avoid overcooking the strawberries. Omit the elderflowers.

Gooseberry and Rhubarb Jam
Use 900g (2lb) gooseberries and 450g (1lb) chopped rhubarb. Cook the fruits separately to avoid overcooking the rhubarb. Omit the elderflowers.

Greengage and Orange Jam

Adding orange greatly improves the flavour of greengage jam.

900g (2lb) greengages
1 large orange
½ lemon
900g (2lb) granulated sugar

Wash the fruit and remove the stones, but reserve them. Put the fruit in a pan and tie the stones in a square of muslin. Add the muslin bag to the pan with the juice and thinly pared and sliced rind of the orange and lemon. Simmer gently for about 10 minutes or until the greengages are soft. Stir in the warmed sugar and continue simmering gently, stirring continuously until the sugar has dissolved and then bring to a rolling boil. Boil for about 10 minutes until setting point is reached (see page 109). Remove from the heat and squeeze the muslin bag. Open it and crack some of the stones with a nutcracker; remove the kernels. Stir these kernels into the jam and then pot and cover in the usual way.

VARIATIONS
Greengage, Orange and Walnut Jam
Stir in 50g (1¾oz) chopped walnuts after setting point is reached.

Greengage and Lemon Jam
Omit the orange and use just the juice of 1 lemon.

Trelissick Kea Plum Jam

Kea plums grow on the banks of the River Fal near Trelissick Garden, a National Trust property in Cornwall. They are a damson variety: small, dark and excellent for jam making. The harvest used to continue throughout August and September 'with all hands available for shaking, picking, sorting and counting'. So many trees overhung the creeks that it used to be possible to shake the plums directly into boats. Ferns were used to line the baskets to protect and preserve the fruit (a trick also used when picking blackberries). Use a good-flavoured plum, preferably home grown, or damsons. You can crack some of the stones and remove the kernels and then add them to the plums to give extra flavour if you wish.

900g (2lb) good-flavoured plums
100ml (3½fl oz) water
Juice of 1 lemon
900g (2lb) granulated sugar

Wash the plums, discarding any leaves and stalks, then halve and stone. Place the plums in a pan with the water and lemon juice. Cook gently for about 10 minutes, until the skins have softened. Stir in the warmed sugar until it has completely dissolved and then increase the heat and bring to the boil. Boil rapidly for about 10 minutes, or until setting point is reached (see page 109) and then pot and cover in the usual way.

If you are using small plums or damsons, cook the fruit whole and remove the stones as they rise to the surface, once the sugar has been added. You can skim them off with a slotted spoon.

VARIATIONS
Kea Plum, Brandy and Walnut Jam
Simmer the plums with 1 small cinnamon stick and continue as above. After setting point is reached, stir in 50g (1¾oz) chopped walnuts and 2 tablespoons brandy. Leave to stand for about 5 minutes and then remove and discard the cinnamon stick. Stir the jam gently and pot as usual.

Kea Plum and Apple Jam
Make as above using half plums and half apples, cooked separately, but then boiled up together with the sugar.

Jellies

Only the juice of fruit is used for jelly making, so the yield is much lower than that of jam. The best fruits are those with a strong natural flavour, such as redcurrants, blackcurrants, blackberries, elderberries and damsons. A good pectin content is essential, so fruits with a good flavour but modest pectin content are usually combined with ingredients that are rich in pectin (see page 108). Cooking apples (windfalls can be used) and crab apples are both excellent for their pectin content. Quinces combine high pectin content with good flavour, and these make successful jellies.

Piquant herb jellies, sharpened with vinegar, are versatile accompaniments to fish, roast meats, poultry and game, cold cuts and raised pork or game pies. They can be based on apple juice, made from cooking apples, windfalls or crab apples, or orange and/or lemon juice. The most successful herb jellies are sage, mint, scented geranium, lemon balm, marjoram, rosemary, oregano, thyme and basil, but do experiment. The herbs can be chopped or added in sprigs.

A good jelly should be clear, sparkling and full of flavour. It should retain its shape when cut, but should not be too stiff to spread. As well as using jellies in all the sweet roles familiar to jam, they can also be used in marinades or for basting grilled and roast meat during the final stages of cooking. Fragrant homemade jellies make very attractive presents.

To Make Jelly

Choosing and preparing the fruit
1. The fruit should never be over-ripe because a good set is achieved when acid, sugar and pectin are present in the correct proportions. There is more acid and pectin in under-ripe fruit.
2. Discard any bruised or damaged fruit. It is not necessary to remove peel, cores and stalks, as they will be extracted when the pulp is strained. Wash the fruit.
3. Cook in water (according to the recipe) with any spices to release the acid and pectin, stirring and mashing frequently with a potato masher, until the fruit is very soft and pulpy.

Straining the juice
1. Scald a jelly bag in boiling water and suspend it over a large, perfectly clean bowl. (Use a jelly bag stand or an upturned stool.)
2. Spoon the fruit pulp into the bag and leave to drain for at least 4 hours, preferably overnight. Let the juice drip naturally (if you squeeze the bag or push the juice through, the jelly will be cloudy). The juice is not clear when strained from the fruit, but it clears on boiling with sugar.

Completing the jelly
1. Measure the strained juice and weigh out the sugar, generally 450g (1lb) sugar for each 600ml (1 pint) juice. Slightly less or more sugar can be added for fruit with lower or higher than average pectin contents.
2. Warm the sugar as for jam (see page 108).
3. Put the juice into a clean, heavy-based pan and heat gently with the sugar, stirring until the sugar has completely dissolved. Boil rapidly until setting point is reached (see page 109).
4. Remove the pan from the heat, and then skim off any scum from the surface with a draining spoon. Any remaining scum can be removed by drawing a piece of kitchen paper across the top. Adding a knob of butter just before skimming helps to disperse the scum.
5. Pour the jelly very gently into warm, sterilised jars. Don't pour too quickly as this will create bubbles in the jelly. Cover and label as for jam (see page 109). Take care not to move the jars while the jelly is setting, or it will split.

Apple Jelly

This is a tasty jelly in its own right, but is also a good base for herb jellies. Use windfall apples if you wish, but choose cookers, or sharper varieties of eating apples, to get a good set.

1·5kg (3lb 5oz) apples
1·2 litres (2 pints) water
Granulated sugar

Wash the fruit and chop roughly, including the peel and cores. Put in a large pan with the water and simmer for about 45 minutes until very soft. Strain through a jelly bag overnight.

Next day, measure the juice and allow 450g (1lb) sugar for every 600ml (1 pint). Place the juice and warmed sugar in a clean pan and heat gently, stirring continuously until the sugar has dissolved and then bring to the boil. Boil rapidly for about 10 minutes, or until setting point is reached (see page 109). Skim, then pot and cover.

VARIATIONS
Apple Ginger Jelly
Place 25–50g (1–1¾oz) bruised dried root ginger in a muslin bag and add to the apples at the beginning of cooking.

Apple Cinnamon Jelly
Place 1 cinnamon stick in a muslin bag and cook with the apples.

Apple Geranium Jelly

Use scented geranium leaves to make this unusual jelly.

6 scented geranium leaves
850ml (1½ pints) apple juice
   (see page 129)
700g (1½lb) granulated sugar

Wash and dry the geranium leaves, then tie them in a piece of muslin. Pour the apple juice into a pan and add the warmed sugar and the muslin bag. Stir over a very gentle heat until the sugar has completely dissolved. Boil rapidly for 5–10 minutes, or until setting point is reached (see page 109). Remove from the heat and discard the muslin bag. Skim, then pot and cover. Leave for a week or two before using to give the geranium flavour time to infuse.

Apple Lemon Jelly

A very good jelly to serve with rich meats such as lamb, pork, duck or goose, but also excellent with oily fish such as mackerel, especially smoked.

1.35kg (3lb) apples
1.2 litres (2 pints) water
2 large thin-skinned lemons

Wash the fruit and chop roughly, including the peel and the cores. Put in a large pan with the water. Using a potato peeler, remove the peel from the lemons without removing any pith, tie in a square of muslin and add to the apples. Continue as for Apple Jelly (see page 130).

Crab Apple and Clove Jelly

Crab apples, still common along roadsides, in hedgerows and on heathland, can be picked from August onwards. They make a superb jelly, which can vary from bright red to yellow depending on variety and ripeness.

This spiced jelly is best eaten with meat or poultry, but if you leave out the cloves the jelly is super as a sponge filling or spread on Scotch pancakes, scones and crumpets.

1·5kg (3lb 5oz) crab apples
Water to cover
Granulated sugar
1 vanilla pod
A few whole cloves

Wash the apples and chop roughly without peeling or coring, into a large pan. Pour in enough water to barely cover and then simmer gently for about 1 hour until the fruit has softened. Strain through a jelly bag overnight.

Next day, measure the juice and allow 450g (1lb) warmed sugar to each 600ml (1 pint). Put the juice and sugar in a pan with the vanilla pod. Heat gently, stirring constantly until the sugar has dissolved, then boil rapidly for about 10 minutes until setting point is reached (see page 109). Skim and remove the vanilla pod and then pour into warm, sterilised jars, adding 2 cloves to each.

Rowan and Crab Apple Jelly

Rowanberries, the fruit of the mountain ash, which is common throughout Britain, are best picked in October. They make the most delicious, tangy, orange-red jelly, which is the traditional and best accompaniment for venison, grouse and hare. I also like to serve it with mutton, lamb or goose. Adding apples improves the set – use cooking apples instead of crab apples if you wish.

900g (2lb) rowanberries
900g (2lb) crab apples
Granulated sugar

Remove the rowanberries from their stalks and wash well. Chop the apples roughly and place both fruits in a pan with just enough water to cover them. Simmer gently for about 45 minutes, or until the fruit is very soft, crushing with a potato masher, then strain through a jelly bag overnight.

Measure the juice and pour into a clean pan with the warmed sugar, allowing 450g (1lb) sugar to each 600ml (1 pint). Heat gently until the sugar has dissolved, then boil rapidly for about 10 minutes until setting point is reached (see page 109). Pot and cover.

VARIATIONS
Sloe and Crab Apple Jelly
Use sloes instead of rowanberries to make a tangy, dark wine-red jelly, especially good with mutton, rabbit or hare.

Elderberry, Bilberry or Cranberry and Crab Apple Jelly
Substitute elderberries, bilberries or cranberries for the rowanberries. This is particularly good with turkey, guinea fowl or pheasant.

Piquant Mint and Apple Jelly

Particularly delicious with lamb and mutton.

1.35kg (3lb) apples
1.2 litres (2 pints) water
2 large thin-skinned lemons
A bunch of fresh mint, plus some extra finely chopped mint
100ml (3½fl oz) white wine vinegar

Follow the recipe for Apple Lemon Jelly (see page 131) and add the bunch of fresh mint, tied loosely in another square of muslin, and the vinegar to the apples when cooking. Continue as for Apple Jelly (see page 130). After setting point has been reached, stir in the finely chopped mint. Leave the jelly to cool a little until a thin skin forms on the surface, then stir again gently to distribute the mint. Pot and cover.

Rosemary, Thyme, Basil or Marjoram Jelly

8–10 small fresh sprigs of your chosen herb
2 large fresh sprigs of your chosen herb
850ml (1½ pints) apple juice (see page 129)
700g (1½lb) granulated sugar

Blanch the small sprigs of herbs in boiling water for 2–3 seconds, rinse under cold water, pat dry with kitchen paper and reserve. Wash the large sprigs of herbs and dry them. Make the jelly as for Mint or Sage Jelly (see page 146), adding the large herb sprigs while dissolving the sugar. Remove the herbs and boil until setting point is reached (see page 109). Put the reserved small herb sprigs into sterilised jars, then pour in the jelly and cover.

Orange and Rosemary Jelly

Very good with lamb and mutton, chicken and pork. Try also with well-flavoured fish, such as salmon, monkfish or turbot.

900g (2lb) oranges
225g (8oz) lemons
1·4 litres (2½ pints) water
1 tablespoon chopped fresh rosemary
Granulated sugar

Wash and halve the oranges and lemons and then slice into semi-circles. Place in a large pan with the water and half the rosemary. Bring to the boil and then simmer gently for about 1 hour, or until the fruit is soft. Strain through a jelly bag and then measure the juice and return to a clean pan, adding 450g (1lb) warmed sugar for every 600ml (1 pint) of liquid. Heat gently, stirring until the sugar has dissolved, then bring to the boil and boil rapidly until setting point is reached (see page 109). Skim and then gently stir in the remaining rosemary. Pour into hot, sterilised jars and cover.

VARIATIONS
Orange and Thyme Jelly
Make as above, but use 2 tablespoons of fresh thyme leaves in place of the rosemary. Serve with poultry and game.

Orange and Sage Jelly
Make as above, but use 2 tablespoons of finely chopped fresh sage leaves in place of the rosemary. Serve with pork and sausages.

Melanie's Bramble Jelly

This is my sister's recipe for a favourite jelly. She recommends that you include some unripe berries to help the set. Cheap and simple to make, it has such a wonderfully evocative flavour. Spread on thick slices of freshly baked bread or scones, or serve with poultry or game birds.

900g (2lb) blackberries
225ml (8fl oz) water
Juice of 1 lemon
400g (14oz) granulated sugar

Wash the blackberries, then place them in a pan with the water and lemon juice. Bring to the boil, then simmer for about 20 minutes until very soft, crushing with a potato masher. Strain the fruit through a jelly bag overnight.

Next day, pour the juice into a clean pan. Add the warmed sugar and stir over a gentle heat until the sugar has completely dissolved. Bring to the boil and boil rapidly for about 7 minutes or until setting point is reached (see page 109). Skim, then pot and cover.

VARIATION
Spiced Bramble Jelly
Tie 1 tablespoon of whole cloves and 1 cinnamon stick, lightly crushed, in a piece of muslin and add to the initial cooking of the blackberries.

Mulberry Jelly

If you are lucky enough to have a mulberry tree in your garden, or can get hold of some of this fabulous fruit, try making this jelly. Mulberries are not rich in pectin, so use with equal quantities of apples to ensure a good set, and do not use very ripe mulberries.

1kg (2¼lb) cooking apples
1kg (2¼lb) mulberries
Water to cover
Granulated sugar

Wash the apples and chop roughly without peeling or coring. Cook the apples and the mulberries separately, with just enough water to cover, until they are really soft and pulpy. Mix the two fruit pulps together and leave to strain through a jelly bag overnight.

Next day, measure the juice and return it to a clean pan, adding 350g (12oz) warmed sugar to each 600ml (1 pint). Heat gently, stirring constantly until the sugar has dissolved, then bring to the boil and boil rapidly until setting point is reached (see page 109). Skim and then pour into warm, sterilised jars and cover.

Damson Jelly

Beatrix Potter made damson jelly every September using damsons gathered from the orchards at Hill Top, her farmhouse in Cumbria. The fruit is so rich in pectin and acid that it invariably makes a very good jelly. Spread on bread, scones, muffins and crumpets, or use as an accompaniment to poultry, ham, lamb or mutton. Beatrix served hers with the local Herdwick lamb or mutton, which is still being raised on the farms she left to the National Trust.

2kg (4½lb) damsons
850ml (1½ pints) water
Granulated sugar

Wash the damsons, discarding any leaves and stalks. Place the damsons in a large pan with the water and simmer for about 40 minutes until very soft, crushing with a potato masher. Strain through a jelly bag overnight. Measure the strained liquid and pour it into a clean pan with 450g (1lb) warmed sugar for each 600ml (1 pint). Heat gently, stirring, until the sugar has dissolved, then boil rapidly until setting point is reached (see page 109). Start testing after 3–4 minutes. Skim, then pot and cover.

VARIATIONS

Spiced Damson Jelly
Tie four whole cloves and 1 cinnamon stick in a piece of muslin and put into the pan with the damsons and water. Continue as above.

Damson and Crab Apple Jelly
Simmer 2kg (4½lb) diced crab apples with 1kg (2¼lb) damsons for about 1½ hours until soft and pulpy. Continue as above.

Blackcurrant Jelly
Blackcurrants are also rich in pectin and acid and make a very good jelly. If you pick the fruit yourself, do not bother to strip them off their stalks – simply use the strings as they are, after washing. Make in the same way as Damson Jelly, but use 1·2 litres (2 pints) water. Particularly good with duck or goose, or use to baste lamb during cooking.

Gooseberry Jelly
Green gooseberries are very rich in pectin and acid, so make an exceptional jelly. Can also be used as a base for herb jellies, instead of apples. Make in the same way as Damson Jelly.

My Mother's Redcurrant Jelly

When redcurrant jelly first appeared in the eighteenth century it was served with venison and hare, rather than lamb. My mother's recipe uses a lot of fruit to produce a small amount of jelly, but it is worth it, as you end up with a 450g (1lb) jar of firm, jewel-like jelly that tastes of redcurrants, rather than just sugar. Eat with game, lamb and mutton, or cheese, or use as the base for Cumberland Sauce (see page 50). Redcurrants are full of pectin-stuffed pips, so setting is easy.

1kg (2¼lb) redcurrants
1 tablespoon water
Granulated sugar

Wash the redcurrants, but don't bother to string them. Place them in a pan with the water and simmer gently for about 1 hour, until the fruit is soft. Mash well with a potato masher and then strain the pulp through a jelly bag for a few hours, or overnight.

Measure the juice and allow 600g (1lb 5oz) warmed sugar for every 600ml (1 pint). Put the juice and sugar into a pan and heat gently, stirring constantly to dissolve the sugar. Bring to boiling point and boil for just a few minutes (it may only take 1 minute, so be careful) to setting point (see page 109). Skim the jelly and then pour immediately into small, sterilised jars and cover. The jelly will keep for one year in a cool, dark place, but eat within one month once the jar is opened and keep it in the fridge.

Variations

Spiced Redcurrant Jelly
Add 3 whole cloves and 1 cinnamon stick, lightly crushed, to the redcurrants.

Redcurrant and Whitecurrant Jelly
Use a combination of redcurrants and whitecurrants to make a milder, less sharp jelly. This is good with roast lamb and for adding to sauces.

Redcurrant and Raspberry Jelly
Use half redcurrants and half raspberries. Cook the redcurrants until soft and then add the raspberries and continue cooking until they are soft. Mash and continue as above, allowing only 450g (1lb) sugar for each 600ml (1 pint) juice. The yield and set for this jelly are better than for raspberry jelly on its own. Loganberries can be used in the same way.

Spiced Plum Jelly

Serve this delicious jelly with pâtés and terrines, or with roast poultry, especially duck and goose.

1·5kg (3lb 5oz) good-flavoured plums
4 slices of fresh root ginger, peeled
2 small cinnamon sticks, halved
2 star anise
300ml (½ pint) water
Granulated sugar

Wash the plums, discarding any leaves and stalks, then place in a large pan with the ginger, spices and water. Simmer, covered, for about 30 minutes or until the plums are soft and pulpy. Strain through a jelly bag overnight.

Next day, measure the juice and add 350–450g (12oz–1lb) warmed sugar to every 600ml (1 pint). The amount of sugar will vary according to how sweet your plums are. Heat gently in a clean pan, stirring constantly until all the sugar has dissolved, then bring to the boil. Boil rapidly for about 10 minutes until setting point is reached (see page 109). Skim and pour into warm, sterilised jars and cover.

Variation
Piquant Plum and Chilli Jelly
Cook the plums as above, then add 150ml (¼ pint) cider vinegar. Boil for 5 minutes, then strain overnight. Continue as above, adding 3 deseeded and sliced medium-hot chillies after the sugar has dissolved. Boil until setting point is reached, then pour into warm, sterilised jars. Leave to stand for about 20 minutes, then stir to distribute the chillies evenly. Pot and cover.

Rhubarb and Orange Jelly

Rhubarb does not have much pectin and has to be mixed with other selected fruits. Orange and lemon are very successful and the jelly would make a good base for herbs. Choose sticks of young, pink rhubarb for the best colour and flavour.

1·5kg (3lb 5oz) rhubarb, trimmed
200ml (7fl oz) water
4 large oranges
Granulated sugar

Wipe the rhubarb and cut into pieces. Place in a pan with the water and the juice, pith and grated rind of the oranges. Simmer gently until the rhubarb is very tender. Strain through a jelly bag overnight.

Next day, measure the juice and allow 450g (1lb) warmed sugar for each 600ml (1 pint). Place in a clean pan and heat gently, stirring continuously until the sugar has dissolved, then bring to the boil. Boil rapidly until setting point is reached (see page 109), then skim. Pour into warm, sterilised jars and cover.

Old-fashioned Quince Jelly

This very fine, deep-pink jelly, made from the true orchard quince (*Cydonia vulgaris*), has been known since Elizabethan days. Quinces are now very fashionable again and their jelly is extremely versatile. It makes a marvellous accompaniment to cheese – equally good with a full-flavoured Cheddar, Stilton or Camembert – to meat, especially lamb, mutton and ham and to all kinds of poultry. Also delicious topped with cream as a filling for sponges, or spread on scones and muffins. Makes a sparkling glaze for fruit flans and cheesecakes. Precious quinces can be eked out by adding a portion of windfall or crab apples.

1·35kg (3lb) ripe quinces
Water to cover
Thinly pared rind and juice of
　1 large lemon
Granulated sugar

Wash the quinces and rub off the soft down on their skins. Do not peel or core them, but chop them roughly into a large pan and just cover with water. Add the lemon rind, then simmer gently for about 1 hour, or until the fruit is very soft and pulpy. Stir in the lemon juice, then strain through a jelly bag overnight.

Next day, measure the juice and pour into a clean pan. Add warmed sugar, allowing 450g (1lb) to each 600ml (1 pint) juice. Heat gently, stirring until the sugar has completely dissolved, and then bring to the boil. Boil rapidly until setting point is reached (see page 109). Skim, then pot and cover.

Variations
Japonica Jelly
Use japonicas (*Chaenomeles japonica*), a closely related species to the quince. Pick when very ripe and sweet-smelling to make a very good jelly – clear, slightly sharp, rather like rowan jelly in flavour, although darker in colour. Marvellous with cheese, game, lamb and mutton.

Medlar Jelly
Substitute medlars for the quinces, but use them before they get to the softened, bletted stage. The jelly is excellent with poultry and game.

Mint or Sage Jelly

15g (½oz) fresh mint or sage leaves
850ml (1½ pints) apple juice (see
    recipe for Apple Jelly, page 130)
700g (1½lb) granulated sugar
Natural green food colouring
    (optional)

Blanch the mint or sage leaves in boiling water for 2–3 seconds, then drain and rinse under cold water. Drain again and pat dry with kitchen paper, then finely chop the herbs (there should be about 1 tablespoon).

Pour the apple juice into a pan, add the warmed sugar and stir over a gentle heat until it has dissolved completely. Boil rapidly for 5–10 minutes or until setting point is reached (see page 109). Remove from the heat, skim, then stir in the chopped mint or sage and a few drops of green colouring if using. Pot and cover.

VARIATION
Herb Garden Jelly
Use a mixture of fresh herbs to make a savoury jelly that goes well with a variety of meats, poultry or fish. Follow the above recipe, but use a combination of fresh parsley, mint, thyme and tarragon. To balance the flavour of the herbs, use more parsley and mint than the stronger thyme and tarragon.

Fruit Butters, Cheeses, Pastilles, Leathers and Curds

These preserves are all basically the same mixture, but are cooked to different consistency with varying proportions of sugar. A large quantity of fruit yields only a relatively small amount of finished preserve, but the flavour is concentrated.

To Make Fruit Butters, Cheeses and Pastilles
1. Simmer the fruit until soft and pulpy.
2. Rub the cooked fruit through a fine nylon sieve to make a pulp, leaving only a debris of pips, stones and skins.
3. Weigh the pulp and put in a pan with the appropriate amount of sugar.
4. Boil rapidly until the required consistency is reached and then pot as for jam or pack in the appropriate way: firm cheeses are packed into wide-topped jars or containers so that they can be unmoulded and sliced for serving. Store in a dry, dark place.

This chapter also includes fruit curds. They are similar to fruit butters only in that they are smooth, with a spreading texture, but they are made from fresh eggs and butter and do not keep for long.

Fruit Butters

These have a soft butter-like consistency and can be used in the same way as jam. They are delicious spread on bread and butter, or served with scones and cream. They also make excellent fillings for sponges and tarts.

Fruit butters are made with 225–350g (8–12oz) sugar to every 450g (1lb) fruit pulp. They should be put into small, warm, sterilised jars or pots and covered as for jam (see page 109). Fruit butters will keep for only three to six months without spoiling as they have a lower sugar content than fruit cheese, so they should be made in small batches.

Bramble and Apple Butter

This is heavenly spread thickly on scones and Cornish or Devonshire splits, with clotted cream. It is also good with a cold roast or as an accompaniment for cream cheese at the end of a meal.

900g (2lb) blackberries
900g (2lb) cooking apples
Grated rind and juice of 2 lemons
Granulated sugar

Wash the fruit and chop the apples roughly, including the peel and cores. Place the blackberries and apples in a pan with the lemon rind and juice, then simmer gently for about 15 minutes until very soft. Rub through a fine nylon sieve and weigh the pulp.

Return the pulp to the pan and add 350g (12oz) warmed sugar to each 450g (1lb) blackberry pulp. Bring slowly to the boil, stirring frequently until the sugar has dissolved. Continue boiling, stirring occasionally, until the mixture thickens and there is no extra liquid left in the pan. Spoon the butter into warm, sterilised jars and cover immediately.

Spiced Apple Butter

This is one of the oldest recipes for a fruit butter. It is very like the nineteenth-century apple 'marmalade' which was said to be 'very nice, and extremely wholesome as supper for the juveniles, and for the aged, eaten with cream or milk'. Spread it on crusty bread and butter or hot buttered muffins.

2·75kg (6lb) apples (windfalls or crab apples can be used)
1·2 litres (2 pints) water
1·2 litres (2 pints) sweet cider
Granulated sugar
½ teaspoon ground allspice
1 teaspoon ground cloves
2 teaspoons ground cinnamon

Wash the apples thoroughly and remove their stalks and any bruised parts. Chop the fruit roughly, place in the pan with the water and cider, and then bring to the boil. Cover the pan and simmer until the apples are soft and pulpy, stirring frequently. Rub the apple pulp through a fine nylon sieve, then weigh it and return to the pan. Allow 350g (12oz) warmed sugar for each 450g (1lb) apple pulp and add to the pan with the spices. Bring gently to the boil, stirring frequently until the sugar has dissolved. Continue boiling, stirring occasionally, until the mixture thickens and there is no extra liquid left in the pan. Put the butter into warm, sterilised jars and cover immediately.

VARIATIONS
Apple Ginger Butter
Instead of the allspice, cloves and cinnamon, use 2 teaspoons ground ginger.

Cidered Apple Butter
Use all cider instead of a mixture of cider and water.

Blackberry, Apple, Orange and Juniper Butter

Spread on toasted muffins or crumpets, or eat with cold roast pork or baked ham.

1kg (2¼lb) cooking apples
600g (1lb 5oz) blackberries
4 oranges
4 lemons
300ml (½ pint) cranberry juice
1 teaspoon juniper berries, crushed
Granulated sugar

Roughly chop the apples (peel, cores and all) and place in a large pan with the blackberries. Remove the peel from the oranges and discard; chop the flesh. Squeeze the juice from the lemons. Add the orange flesh, lemon juice and lemon skins to the pan. Pour over the cranberry juice and add the juniper berries. Bring gently to the boil and simmer until the fruit is soft and pulpy. Remove the lemon skins and leave to cool. Push the pulp through a nylon sieve and weigh. Place in a pan with the sugar, allowing 350g (12oz) warmed sugar to every 450g (1lb) pulp. Bring slowly to the boil, stirring frequently until the sugar has dissolved. Continue boiling, stirring occasionally until the mixture thickens and there is no extra liquid left in the pan. Spoon the butter into warm, sterilised jars and cover immediately

Plum Butter

Any variety of plum makes good butter, particularly dark-skinned fruit. A similar butter can be made using damsons. As well as being superb in the usual sweet-preserve roles, they can be used as a standby for making plum sauce in Chinese cookery.

1·8kg (4lb) ripe good-flavoured plums
Granulated sugar

Wash the plums and slit the skins. Put in a pan with a little water and simmer gently for about 20 minutes, or until very soft. Press through a nylon sieve until only the skins and stones are left. Weigh the pulp and return to the pan with the appropriate amount of sugar, allowing 350g (12oz) warmed sugar to each 450g (1lb) plum pulp. Continue as for Bramble and Apple Butter (see page 150).

Rhubarb and Orange Butter

4 oranges
1·8kg (4lb) rhubarb, trimmed and
    roughly chopped
Granulated sugar

Peel the oranges thinly with a sharp knife or potato peeler, removing only the zest and none of the pith. Cut the oranges in half and squeeze out the juice. Place the rhubarb in a pan with the peel and juice of the oranges. Add just enough water to cover the fruit. Bring to the boil, cover the pan and simmer until the rhubarb is very tender. Rub the pulp through a fine nylon sieve. Weigh the pulp and return to the pan with 225g (8oz) sugar for each 450g (1lb) pulp. Continue as for Bramble and Apple Butter (see page 150).

Fruit Cheeses

These are much thicker than fruit butters. They should be cooked until the mixture is so thick that if a spoon is drawn across the base of the pan, it will leave a definite path. Pot in small straight-sided jars or moulds (small ramekin dishes are ideal) so that the cheese can be turned out whole. Before potting, brush the inside of the warmed jars or moulds with a little neutral-flavoured oil (sweet almond or grapeseed, for example) or glycerine to make unmoulding easier. Pour in the hot cheese and cover as for jam (see page 109).

A fruit cheese should be kept for at least two months before opening but, once it has been turned out, it should be used up as quickly as possible (in a pot, they will keep for at least two years, improving all the time). Fruit cheeses are usually sliced or cut into wedges and eaten with meat, poultry (particularly smoked), game, and raised pies or with dairy cheeses.

Crab Apple Cheese

Crab apple or apple cheese was a feature of Victorian dinner tables, especially at Christmas time, when it was eaten as a dessert decorated with whole hazelnuts and whipped cream. The best is made with one type of apple only.

Eat the cheese as a dessert with thick cream poured over it. As an accompaniment to cold roast pork, ham, goose, duck or game, turn out whole and slice at the table.

1·5kg (3lb 5oz) crab apples or windfall
    apples
Water to cover
Granulated sugar

Wash the apples thoroughly and remove the stalks, leaves and any bruised parts. Chop the fruit roughly across the core to expose the pips. (This is important for the flavour of the cheese.) Put the apples in a large pan with enough water to just cover. Bring to the boil, cover the pan and simmer until the apples are soft and pulpy. Rub through a fine nylon sieve. Weigh the sieved apple pulp and measure out 450g (1lb) sugar for each 450g (1lb) pulp. Put the apple pulp in a clean pan and bring to the boil. Boil uncovered, until a thick creamy consistency. Stir in the sugar and gently return the mixture to the boil, stirring continuously until all the sugar has dissolved.

Continue boiling, stirring regularly to avoid sticking, until the mixture is so thick that when you draw a wooden spoon across the bottom of the pan, it leaves a clean line. Pour into warm, sterilised jars or moulds and cover. Store in a cool, dark place for one to two months before using.

VARIATIONS
Spiced Crab Apple Cheese
Add ½ teaspoon ground cloves and ½ teaspoon ground cinnamon to the apple pulp.

Apple and Blackberry Cheese
Substitute 450g (1lb) blackberries for 450g (1lb) apples. Serve with cold pheasant or other game.

Apple and Plum Cheese
Substitute 450g (1lb) plums for 450g (1lb) apples. Serve with hot or cold roast lamb or duck.

Apple and Mint Cheese
Make as for Crab Apple Cheese, but add 4 tablespoons chopped fresh mint just before it is ready. Cook for a further 5 minutes and pot as above. Serve with hot or cold roast lamb.

Fruit Butters and Cheeses

Spiced Cranberry and Apple Cheese

The cranberry used to flourish in the East Anglian Fens before these were drained, but now is very rare, growing only in the north of England, Scotland and Wales. Cranberry cheese is good with cold poultry and game, especially at Christmas. The cheese looks good turned out and left whole, then decorated with orange slices.

450g (1lb) cranberries
450g (1lb) cooking apples
Grated rind and juice of 2 oranges
1 blade of mace
½ teaspoon whole cloves
½ cinnamon stick
300ml (½ pint) water
Granulated sugar

Wash the cranberries and wash and chop the apples, including their peel and cores. Place in a pan with the orange rind and juice and the spices. Pour in the water and cover the pan. Bring to the boil, then simmer for about 30 minutes, or until the fruit is soft and pulpy. Rub through a nylon sieve and weigh the pulp. Place in a clean pan and add 450g (1lb) sugar for every 450g (1lb) fruit pulp. Heat gently, stirring until the sugar has dissolved, and then bring to the boil and continue to boil, stirring frequently, until the mixture is so thick that a wooden spoon leaves a clean line when drawn across the bottom of the pan.

Pour into hot, sterilised jars or moulds and cover. Store in a cool, dry place for one to two months before using.

Damson Cheese

This cheese is one of the oldest traditional country dishes, always found piled up on the shelves of a country store cupboard. The cheese, if properly made, is a dark purple, almost black, and should keep for years. It is at its best when it has shrunk a little from the sides of the jars and the top has begun to crust with sugar.

In the old days, damson cheeses were often poured into deep dinner plates or large shallow pots and stored for several days in a dry, dark cupboard. They were then turned out, stacked one on top of each other with bay leaves in between and covered with muslin or cheesecloth. They were kept like this until crusted with sugar and then brought out for dessert at a special dinner party, often at Christmas, when they were studded with almonds and served with a little port wine poured over.

1·35kg (3lb) damsons
150ml (¼ pint) water
Granulated sugar

Wash the damsons, discarding any leaves or stalks. Put the fruit into a large pan with the water. Cover the pan and simmer gently for about 15 minutes, or until the fruit is really soft and pulpy. Scoop out the stones with a slotted spoon and discard, and then press the damson pulp through a nylon sieve. Weigh the pulp and pour into a clean pan, adding 450g (1lb) sugar for each 450g (1lb) damson pulp.

Heat gently, stirring until the sugar has dissolved, then bring to the boil and continue to boil, stirring frequently, for about 40 minutes, or until the mixture is so thick that a wooden spoon leaves a clean line when drawn across the bottom of the pan.

Pot and cover as usual (see page 109), then store in a cool, dry place for about two months to mature. Eat with cold meats, pâtes and terrines, raised pies and cheese.

VARIATION
Damson and Mint Cheese

Add 3 tablespoons chopped fresh mint just before the cheese is cooked. Cook for a further 5 minutes and pot as above. Serve with hot or cold roast pork or lamb or with game.

Green Gooseberry Cheese

Apparently this cheese was popular in Jane Austen's household. It is very good served with smoked fish, especially mackerel, or with cold meats, particularly lamb, goose or smoked duck. It also makes a refreshing and summery dessert, coated with pouring cream.

1·5kg (3lb 5oz) green gooseberries
300ml (½ pint) water
Granulated sugar

Wash the gooseberries, removing the stalks and leaves, and then cook gently in a large pan with the water for about 30 minutes, or until very soft and pulpy. Rub through a fine nylon sieve. Weigh the pulp and measure out 450g (1lb) sugar for each 450g (1lb) pulp.

Return the gooseberry pulp to a clean pan, bring to the boil and boil, uncovered, until creamy. Stir in the sugar and cook gently, stirring continuously, until the sugar has completely dissolved. Bring back to the boil and continue to boil for about 1 hour or until very thick, stirring frequently. It is ready when a wooden spoon drawn across the bottom of the pan leaves a clean line.

Pour into hot, sterilised jars or moulds and cover. Store in a cool, dry place for one to two months before using.

Quince Cheese

The scented honey flavour of this cheese goes particularly well with crusty bread and both hard and soft cheeses. It is also delicious with cold meat, pâtes, terrines and foie gras.

1.5kg (3lb 5oz) quinces
Water to cover
Granulated sugar

Wash the quinces and rub off any fluff on the skin. Discard any leaves or stalks. Chop roughly across the core and place in a large pan with enough water to barely cover the fruit. Bring to the boil and then simmer, covered, for about 1 hour, or until the quinces are very soft and pulpy. Rub through a fine nylon sieve. Weigh the pulp and measure out 450g (1lb) sugar for each 450g (1lb) quince pulp.

Put the pulp in a clean pan and bring to the boil. Continue boiling, uncovered, until a thick, creamy consistency. Stir in the sugar and gently return the mixture to the boil, stirring continuously until all the sugar has dissolved. Continue boiling for about 1 hour, stirring regularly to avoid sticking, until the mixture is so thick that when you draw a wooden spoon across the bottom of the pan, it leaves a clean line.

Pour into warm, sterilised jars or moulds and cover. Store in a cool; dry place for at least two months before eating, although as it ages, quince cheese becomes firmer, darker and more intense in flavour.

Fruit Leathers

Old-fashioned fruit leather is essentially concentrated fruit purée, dried and rolled into sheets. It has been made for centuries to be eaten when fruit was out of season. The colours are vivid and the flavours intense. Today's supermarkets sell a poor substitute, so make the original ones and know that your children are snacking on something healthy, as well as delicious.

Fruit leather will keep in a cool dark place for two months, or in the fridge for four months, or you can freeze it for up to one year.

Strawberry and Raspberry Leather

100g (3½oz) raspberries
200g (7oz) strawberries
Juice of ½ lemon
50g (1¾oz) honey

Cook the raspberries in a pan until soft and then push through a fine nylon sieve. Discard the seeds and place the raspberry purée back in the pan. Add the strawberries, the lemon juice and the honey and then simmer for 5 minutes.

Blend in a blender or food processor until smooth. Line a medium baking tray with aluminium foil and pour in the fruit mixture. Spread until it just runs to the sides.

Dry in the oven at 70°C, 160°F, gas mark ¼, for about 6 hours. The leather is ready when it is slightly tacky, but not sticky.

Remove from the oven and leave to cool, then roll up the leather in clingfilm, or cut into strips. Store in an airtight container in a cool place, where it will keep for two months – that is if you can hide it from the kids.

Variations
Try other fruits and combinations of fruits, such as black and redcurrants, blackberries, blueberries, plums and mango.

Fruit Pastilles

Apple, plum, apricot and quince pastilles or 'comfits' (rather like today's fruit pastilles) were popular sweetmeats in Tudor and Stuart times. They were cut into shapes, then rolled in sugar and kept in special cupboards for use during the winter when fresh fruit was scarce.

The mixture for fruit pastilles is the same as for a fruit cheese, but the cooking is continued, stirring all the time, until almost dry. The cold mixture will be about the consistency of soft marzipan. Shape the pastille into balls, or oblongs, and wrap in waxed paper to be served as confectionery, or place in paper sweet cases in a pretty box as an unusual gift.

Quince Comfits

450g (1lb) very ripe quinces
150ml (¼ pint) medium-sweet wine
Granulated sugar
A little ground cinnamon
A little ground ginger
A little black pepper
Caster sugar to finish

Quinces are often to be seen growing in National Trust gardens or orchards. Attingham Park, near Shrewsbury, produces fine quinces in a good year and Oxburgh Hall near Norfolk has a quince orchard.

Chop the quinces roughly, including the skin and cores. Simmer with the wine until very tender, then rub through a nylon sieve. Weigh the pulp and return to a clean pan with an equal amount of sugar. Season to taste with the spices, then bring slowly to the boil, stirring frequently until the sugar has dissolved. Cook gently for about 1½ hours, until the mixture is very thick indeed and leaving the side of the pan. Stir frequently to prevent it from burning. Remove from the heat, then leave to cool completely.

Cut into small pieces and roll in caster sugar. Store in a wooden or cardboard box lined, with waxed paper with extra caster sugar to prevent the sweets from sticking together.

Fruit Butters and Cheeses

Townend Apricot Pastilles

This recipe is taken from a commonplace book written in 1699 by Elizabeth Birkett of Townend, a National Trust property near Windermere in Cumbria. She was the wife of a wealthy yeoman farmer, a member of the Browne family who lived in the house from 1626 until 1944. The original recipe mixes a pound of apricot pulp with half a pound of apple pulp to bulk out an expensive fruit, but I think the flavour is better using just apricots.

300ml (½ pint) apricot purée
225g (8oz) granulated sugar
1 tablespoon lemon juice
Caster sugar to finish

Make the purée from about 450g (1lb) fresh apricots, or from 225g (8oz) dried fruit, soaked and cooked.

Put the apricot purée, sugar and lemon juice into a heavy-based pan over a gentle heat, stirring until the sugar has dissolved, then bring the mixture to the boil and cook, stirring frequently, for about 40 minutes, or until a little will set firmly on a cold surface.

Wet a 20cm x 15cm (8 x 6in) tin and pour in the mixture. Leave in a warm room for two to four days, then cut into pieces and roll in caster sugar. Arrange in paper sweet cases in an airtight jar or other container. Makes a very special gift.

Variations
Use peaches, raspberries or blackcurrants in the same way.

Fruit Curds

These contain butter and eggs, so they have a limited shelf life and should be kept in the fridge for up to six weeks. Even when stored in the fridge, it is a good idea to inspect them occasionally for signs of mould. Fruit curds were originally made in stone pots standing in a pan of hot water, then stored in these pots on the still-room or store cupboard shelf. The nearest modern equivalent to this method is to make the curd in a slow cooker, which is really easy (see Trelissick Lemon Curd, below). Alternatively, fruit curds can be made very successfully in a microwave, in a basin over a pan of hot water, or in a double boiler. You do have to keep an eye on them in case they curdle. If this has just begun to happen, remove quickly from the heat and stand in a bowl of cold water. Whisk hard until the curdling has disappeared. Continue to heat until thick enough to coat the back of a wooden spoon and hold a light ribbon trail. Don't overcook the curd as it will thicken as it cools. A good curd should have a smooth consistency and a fresh flavour.

Trelissick Lemon Curd

The restaurant at Trelissick, a National Trust property near Falmouth in Cornwall, makes this delicious lemon curd as a filling for a much-loved homemade lemon sponge. They make it in a slow cooker and assure me it also freezes well.

Finely grated rind and juice of
   2 large lemons
4 large eggs, lightly beaten
175g (6oz) caster sugar
100g (3½oz) unsalted butter,
   melted

Place the lemon rind and juice in a basin or dish that will fit in the slow cooker. Add the beaten eggs, sugar and melted butter and stir well. Cover with a piece of foil and place in the slow cooker. Pour in enough boiling water to come halfway up the basin or dish, then cover with the lid of the slow cooker. Cook for about 1½ hours until thick. Stir well with a wooden spoon, then pot in warm jars. If you do not have a slow cooker, follow the method for Mayfield Apple Curd (see page 166).

Mayfield Apple Curd

This makes a delectable filling for tarts and sponge cakes, especially with whipped cream. It is also good spread thickly on fresh bread and scones, which is how we first enjoyed this particular curd in a tiny tea shop in the village of Mayfield in Sussex.

450g (1lb) cooking apples, cooked and puréed
Finely grated rind and juice of 2 large lemons
4 large eggs, well beaten
450g (1lb) caster sugar
100g (3½oz) unsalted butter, melted

Put the apple purée, lemon rind and juice into a double boiler or a heatproof basin set over a pan of barely simmering water. Take care not to let the bottom of the basin touch the water. Add the well-beaten eggs, sugar and melted butter. Stir the mixture frequently with a wooden or plastic spoon (metal implements can spoil the flavour) for about 20 minutes until thick.

Remove from the heat and pour into warm, sterlised jars. Cover immediately with waxed paper discs and leave until completely cold. Then cover with cellophane, label and store in the fridge for up to six weeks.

Variation
Spicy Apple Curd
Add 1 teaspoon ground cinnamon and 1 teaspoon ground ginger to the apples when cooking. Proceed as above.

Blackberry Curd

This is delicious spread thickly on drop scones, crusty bread or hot buttered muffins.

350g (12oz) blackberries
225g (8oz) cooking apples
Grated rind and juice of 1 large lemon
4 large eggs, well beaten
350g (12oz) caster sugar
100g (3½oz) butter

Wash the blackberries and peel, core and chop the apples. Place in a pan and cook gently together for about 15 minutes or until really soft. Rub through a nylon sieve, then put this purée with the lemon rind and juice into a double boiler or a heatproof basin set over a pan of barely simmering water. Add the well-beaten eggs, sugar and melted butter. Stir the mixture frequently for about 20 minutes until thick.

Remove from the heat and pour into warm, sterilised jars. Cover immediately with waxed paper and leave until completely cold. Then cover with cellophane, label and store in the fridge for up to six weeks.

VARIATIONS
Raspberry or Loganberry Curd
Substitute raspberries or loganberries for the blackberries.

Gooseberry Curd

This curd has a very subtle flavour and makes a delightful spread for fresh bread and butter. It is also good as a filling for tarts, cakes, especially chocolate-flavoured, and as a topping for cheesecakes.

450g (1lb) green gooseberries
About 1 tablespoon water
100g (3½oz) unsalted butter, melted
225g (8oz) caster sugar
4 large eggs, well beaten

Wash the gooseberries and top and tail them. Cook gently in a little water for about 15 minutes or until very soft. Rub through a nylon sieve, then put this purée into a double boiler or a heatproof basin set over a pan of barely simmering water. Add the well-beaten eggs, sugar and melted butter. Stir the mixture frequently for about 20 minutes or until thick.

Remove from the heat and pour into warm, sterilised jars. Cover immediately with waxed paper discs and leave until completely cold. Then cover with cellophane, label and store in the fridge for up to six weeks.

Strawberry and Orange Curd

This curd makes a mouth-watering filling for a Victoria sandwich, with or without cream. If you pick your own strawberries from a farm, this is an ideal recipe for using up any slightly squashed berries at the bottom of the container.

225g (8oz) strawberries
Finely grated rind and juice of 1 large
　orange
4 large eggs, well beaten
225g (8oz) caster sugar
100g (3½oz) unsalted butter, melted

Hull the strawberries and make sure that they are clean, but don't wash them. Mash with a fork, then place with the orange rind and juice in a double boiler or heatproof basin set over a pan of barely simmering water. Add the well-beaten eggs, sugar and melted butter. Stir the mixture frequently for about 20 minutes until thick.

Remove from the heat and pour into warm, sterlised jars. Cover immediately with waxed paper discs and leave until completely cold. Then cover with cellophane, label and store in the fridge for up to six weeks.

Marmalades

The term marmalade dates back to the Middle Ages, but was not originally applied to a preserve of citrus fruits. Instead, it was a stiff paste of quinces, honey and spices, cut into small shapes and served as a sweetmeat (see Fruit Pastilles, page 161). It was a luxury article imported in boxes, usually from Portugal, and was called after the Portuguese for quince, *marmelo*. It became subject to high customs duty, so English housewives learned to make their own marmalade from home-grown quinces.

Gradually 'marmalade' was used to describe any fruit cooked to a pulp with sugar or honey and spices and used to fill pies or tarts or eaten as sweetmeats. Citrus fruits were sometimes included. Oranges and lemons first arrived as luxury items in England in the thirteenth century, but from the end of the fourteenth century the shipments became more frequent and prices fell. The oranges were always of the Seville type, coming in from Spain or Portugal or on the spice ships belonging to the Venetians and the Genoese. In addition to citrus fruits, jars of 'citrenade' were imported. This was a kind of 'marmalade' made from lemons – it was solid in texture and eaten in pieces rather than used as a spread.

Sweet oranges were first brought back from Ceylon during the sixteenth century by the Portuguese and soon spread into the orange-growing countries of southern Europe. Later that century, a superior fruit with a sweet flavour, the 'China' orange, became so fashionable that a few wealthy garden owners, such as Elizabeth I's chief minister, Lord Burghley, tried to coax orange trees to fruit in the English climate.

By the end of the eighteenth century, oranges and lemons had become the most easily grown exotic fruits; elegant buildings to house the tender trees graced many a country-house garden. The National Trust owns many fine orangeries – notably at Dyrham Park, near Bath, at Saltram in Devon and at Felbrigg and Blickling Halls, which are both in Norfolk.

The first true orange marmalade was probably made in the eighteenth century by the Keiller family in Scotland. It is said that James Keiller, a Dundee grocer, was tempted to buy a large quantity of cheap Spanish oranges, which did not sell well in his shop as they were of the bitter Seville type. He took them home and his wife made them into jam, or the first orange marmalade. She had already established a good trade for her quince preserve known as 'marmalet' and the same customers flocked to buy her new product. With her son she set up a factory to make marmalade, which is still thriving today. Alexander Keiller, known as the 'marmalade millionaire', used his marmalade fortune to buy, excavate and re-erect much of the stone circle at Avebury in Wiltshire, now looked after by the National Trust.

Marmalade was not properly introduced into England until 1870 when a grocer's wife from Oxford, a Mrs Cooper, made a small amount. She was overwhelmed by the demand, and marmalade in time became a standard item on the breakfast table.

Although marmalade is a traditional breakfast preserve, it is excellent in cakes, puddings and tarts. Try it also as a filling with cream for an orange or lemon sponge, or as a flavouring for ice cream. Marmalade is very good served with cold ham, pork, duck or goose, or used for basting these meats.

Homemade marmalade is not only more delicious, but also much cheaper than shop-bought. It can vary from a thick, chunky consistency to a clear sparkling jelly, depending on the recipe.

To Make Marmalade

1. Scrub the fruit to remove dirt and chemicals and then prepare according to the recipe. If possible, buy unwaxed or organic citrus fruit. (Cut up the fruit by hand if you want the best, most even result, or use a marmalade cutter or food processor.)
2. Simmer the cut peel in a heavy-based pan, uncovered, until the liquid has reduced by half and the peel is very tender. (It will not become any softer after the sugar has been added.)
3. Warmed sugar dissolves faster when added to the simmered peel (warm it in the oven when you sterlise your jars). Ensure that the sugar has dissolved completely before boiling for a set or it may crystallise later in the preserve. (You will know the sugar has dissolved when the marmalade becomes translucent.)
4. To test for setting, take the pan off the heat and drop a little marmalade on a cold plate. Leave for a few minutes, then push your fingertip through the marmalade. If it wrinkles, it will set. If the marmalade is not ready, put the pan back on the heat to boil for a few minutes longer and test again. (Always remember to take the pan off the heat during testing because over-boiling will ruin your marmalade).
5. After setting point is reached, turn off the heat. Skim with a slotted spoon, to prevent scum adhering to the peel, stir well and let stand for about 15 minutes, to stop the peel rising after potting.
6. Stir gently and then ladle into sterilised jars. Seal at once with waxed paper discs and covers.
7. Allow to cool completely, then label and store in a cool, dark cupboard until required.

Grapefruit and Lemon Marmalade

When Seville oranges are unavailable, this makes a good substitute. Choose thin-skinned grapefruit if possible.

900g (2lb) grapefruit
450g (1lb) lemons
3·6 litres (6 pints) water
2·75kg (6lb) granulated sugar
A knob of butter

Wash the fruit and then thinly pare off the rind with a potato peeler and cut into thin strips. Place in a large pan. Cut all the pith from the fruit with a sharp knife and reserve it. Slice the fruit and add it to the pan with any juice. Tie the pips and reserved pith in a square of double-layered and add to the pan with the water. Bring to the boil, then reduce the heat and simmer gently for 1–1½ hours, or until the peel is very soft and disintegrates when squeezed.

Remove the muslin bag, squeezing the juice back into the pan, and discard. Add the warmed sugar and stir over a low heat until dissolved, then boil rapidly, stirring frequently until setting point is reached (see page 173 – start testing after about 10 minutes). Stir in a knob of butter to disperse the scum. Allow the marmalade to cool for about 15 minutes and then stir. Pot and cover in the usual way (see page 173).

Lemon Shred Marmalade

The pith and pips of lemons are full of natural pectin, so this marmalade sets splendidly in about 5 minutes. Limes can be used in the same way, although setting will take longer.

700g (1½lb) lemons
2 litres (3½ pints) water
1·5kg (3lb 5oz) granulated sugar

Wash the lemons and then pare off the zest with a potato peeler or sharp knife. Cut into very fine 2.5cm (1in) long strips and place in a large bowl.

Cut away all the pith from the lemons and put this in another large bowl. Cut the lemons into segments, cutting away the segment skin and reserving, together with the pips. Add the lemon segments to the bowl containing the strips of zest. Add the segment skin, the pips and the central core of the lemon to the bowl containing the pith.

Add 1 litre (1¾ pints) of cold water to the bowl containing the lemon zest and segments and 1 litre (1¾ pints) of cold water to the bowl of pith, pips etc. Cover both bowls with a clean cloth and leave to steep overnight or for 24 hours.

Next day, put the soaked pith, pips and water in a pan and simmer gently for 1½ hours. Transfer the contents of the pan to a fine sieve, set over a bowl and allow to drain well. Push the pith down with the back of a wooden spoon to extract the remaining pectin liquor and reserve.

Transfer the lemon segments, zest and water to a large pan and simmer for about 1½ hours, or until the peel is very tender and the liquid has reduced by half.

Add the reserved pectin liquor and the warmed sugar and simmer over a low heat, stirring regularly until all the sugar has dissolved. Then bring to the boil and boil rapidly for about 5 minutes, until setting point is reached (see page 173).

Remove from the heat, skim, stir well and leave to stand for 30 minutes for the fruit to settle. Pot and cover in the usual way (see page 173).

Tangerine Marmalade

Tangerines must be combined with other citrus fruits, such as lemons or grapefruit, when making marmalade to improve the pectin content.

800g (1lb 12oz) tangerines
450g (1lb) lemons
3 litres (5 pints) water
1·35kg (3lb) granulated sugar

Wash the fruit thoroughly and remove the stalk ends. Peel the tangerines, cut the peel into very fine strips and tie it loosely in a square of double-layered muslin.

Halve the lemons and squeeze out the juice. Chop up the remaining lemon shells and tie loosely in another muslin bag with all the pips.

Chop up the tangerines and place in a pan with the bag of tangerine peel, the bag of lemon skins, the lemon juice and the water. Bring to the boil and simmer for 30 minutes. Remove the bag of tangerine peel and reserve. Continue cooking gently for a further 1½ hours and then remove the muslin bag of lemon skins, squeezing any liquid back into the pan.

Open the bag of tangerine peel and add the pieces to the cooking pan. Add the warmed sugar and bring slowly to the boil, gently stirring until all the sugar has dissolved. Boil rapidly until setting point is reached (see page 173).

Remove from the heat, skim and stir and then leave to cool for 15 minutes. Pot and cover in the usual way (see page 173).

Three Fruit Marmalade

This is a very useful standby recipe, which does not require seasonal Seville oranges. Follow the suggested proportions of different fruit for a good set, as the lemons are the main source of pectin, rather than the grapefruit or sweet oranges.

1·35kg (3lb) mixed citrus fruit
    (about 2 grapefruit, 2 sweet oranges
    and 4 lemons)
3·6 litres (6 pints) water
2·75kg (6lb) granulated sugar
A knob of butter

Cut all the fruit in half, then squeeze out the juice and pour into a large pan. Tie the pips in a piece of muslin and add to the juice in the pan. Cut the squeezed orange and lemon halves in half again and the grapefruit halves into quarters. Slice the pieces of peel into thick or thin shreds, as you prefer, without removing the pith, and add to the pan with the water. Bring to the boil, then reduce the heat and simmer for 1½ hours, or until the peel is very soft.

Remove the muslin bag, squeezing the juice back into the pan. Add the warmed sugar and stir over a low heat until completely dissolved, then bring to the boil and boil rapidly until setting point is reached (see page 173). Stir in a knob of butter to disperse the scum and then leave to cool slightly. Pot and cover in the usual way (see page 173).

Daddy's Seville Orange Marmalade

Bitter oranges are the most popular fruit for marmalade because of their flavour and appearance. The peel of sweet oranges gives marmalade a rather cloudy finish and the pith does not turn as translucent. Unfortunately, Sevilles and other bitter oranges have a very short season – usually just before Christmas to the first week of February. They can be frozen for use later in the year, although my father, who was the 'marmalade king' in our family, thought that the resulting marmalade was inferior. Here is his recipe, which is excellent and frequently won prizes in our local produce show.

If you want a perfect jellied set, buy your Seville oranges as soon as you see them in the shops, because they will contain more pectin than the later ones. If you are using Sevilles at the end of the season, substitute preserving sugar for half the weight of granulated sugar, for the firmest set, and add extra oranges.

900g (2lb) bitter oranges,
　　preferably Seville oranges
1 sweet orange and 1 lemon
2·4 litres (4 pints) water
1·8kg (4lb) granulated sugar
A knob of butter

The skins of Seville oranges are often quite dirty, so wash them well before cutting them in half and removing and reserving all the pips. Shred all the fruit fairly finely, including the pith. Steep the reserved pips in a small basin with 600ml (1 pint) of the water. Put the shredded fruit in a large bowl and pour in the remaining water. Cover both bowls with a clean cloth and leave to stand overnight.

Place the fruit and its liquor in a large pan and bring to the boil. Boil for 1½–2 hours or until the peel is very soft and disintegrates when squeezed. (The peel should not be overcooked, as this will spoil the colour of the marmalade). Strain the water from the reserved pips into the pan of fruit and tie the pips in a square of double-layered muslin.

Add the muslin bag to the pan and gradually stir in the warmed sugar. Heat gently until the sugar has completely dissolved, then increase the heat and bring the mixture back to the boil. Boil briskly, stirring frequently to prevent catching on the bottom of the pan, for up to 30 minutes, or until setting point is reached (see page 173 – start testing after 10 minutes). Discard the muslin bag. Add a knob of butter to disperse any scum and leave to stand for about 10 minutes. Stir, then pot and cover in the usual way (see page 173).

Variations
Ginger Marmalade
Follow the above recipe, using Demerara sugar. Stir in 100g (3½oz) preserved stem ginger, finely chopped, with the warmed sugar. Continue as before. Grapefruit also makes a good base for Ginger Marmalade.

Dark Oxford Marmalade
This is a chunky, coarse-cut marmalade for true marmalade lovers. Its rich, dark colour is achieved by adding black treacle. Follow the main recipe above, cutting the fruit into thick shreds or chunks. Add 1 tablespoon black treacle with the sugar. Store for several months before opening.

Dark Oxford Marmalade with Whisky
Make as above and stir in 2 tablespoons whisky after setting point has been reached. Store for several months before opening.

Marmalades 179

Seville Orange Jelly Marmalade

My father used to make this especially for me, as I really prefer fine jelly marmalade. It is very good served as an accompaniment to cold ham, pork, duck or even pork sausages, as well as being delicious spread thickly on warm buttered toast. If a completely clear jelly is required, don't remove any peel from the oranges. Other citrus fruits can be used to make jelly marmalades.

900g (2lb) Seville oranges, peeled and chopped
1 sweet orange, peeled and chopped
2 lemons, peeled and chopped
4·2 litres (7 pints) water
1·8kg (4lb) granulated sugar

If you want some peel suspended in the jelly marmalade, thinly pare the rind from half the Seville oranges. Shred the rest of the rind very finely and steep in 300ml (½ pint) water in a large bowl. Cover the chopped fruit with 1·7 litres (3 pints) water; leave to steep for 24 hours.

Turn the fruit and steeping liquor into a large pan and add the remaining water. Strain the shredded peel and add its soaking liquor to the pan. Tie the peel in a piece of muslin and add it. Bring the fruit to the boil and simmer gently for 1¼ hours. Remove and reserve the muslin bag of shredded peel, gently pressing out excess liquid back into the pan. Continue to cook the fruit for a further 15–45 minutes or until the mixture has reduced by half. Remove the peel from the muslin and rinse in cold water, then drain. Strain the fruit mixture through a jelly bag and set aside in a covered container.

Measure the strained liquid and pour it back into the pan. Add 450g (1lb) warmed sugar for each 600ml (1 pint) liquid. Add the reserved shredded peel and stir well over a gentle heat until the sugar has dissolved completely. Increase the heat and bring to the boil. Boil rapidly for about 20 minutes or until setting point is reached (see page 173 – start testing for setting point after 15 minutes). Skim, then allow the marmalade to stand for about 10 minutes. Stir gently to distribute the peel evenly, then pot and cover in the usual way (see page 173).

Dark Coarse-cut Marmalade

The true Oxford type of marmalade that appeals to real marmalade lovers needs to be stored for several months to bring out its full flavour.

1.5kg (3lb 5oz) Seville oranges
Juice of 2 lemons
3.6 litres (6 pints) cold water
2.75kg (6lb) granulated sugar
1 tablespoon black treacle

Wash the oranges well, discarding any stalks and leaves, and cut in half. Squeeze out the juice and reserve the pips. Cut the peel into thick shreds or chunks, reserving excess pith. Chop the flesh roughly and place in a large pan with the peel. Tie the reserved pips and pith in a square of muslin and add to the pan. Pour in the lemon juice and water, then bring slowly to the boil, stirring occasionally. Reduce the heat and simmer gently for about 2 hours, or until the peel is very tender. Remove the muslin bag and give it a good squeeze, returning any liquor to the pan.

Stir in the warmed sugar and treacle and slowly bring to the boil, stirring continuously until the sugar has completely dissolved. Boil rapidly for about 15 minutes, or until setting point is reached (see page 173). Remove from the heat, skim, then allow to cool a little before pouring into warm, sterilised jars. Cover as usual and store in a cool dark place for at least one month before eating.

Lime Marmalade

Although Lime Marmalade – my favourite – can be made using the recipe for Lemon Shred Marmalade (see page 175), I prefer this recipe. Limes and grapefruit arrived in Britain later than oranges and lemons. By the 1680s, casks of lime juice were being exported from Jamaica to make punch, a drink newly popular in England. Later, the British Navy, realising that lime juice helped to fight scurvy, made it compulsory for their sailors – hence the nickname of 'Limey' for a Briton.

700g (1½lb) limes
1·75 litres (3 pints) water
1·35kg (3lb) granulated sugar

Wash the limes and remove the stalk ends. Place them in a large pan with the water and simmer gently for about 2 hours or until the fruit is very soft. Remove the limes and slice very thinly, discarding the pips. Put the fruit and any excess juice that has collected during slicing into the pan with the original water and then add the warmed sugar. Simmer gently, stirring continuously until all the sugar has dissolved, then boil rapidly until setting point is reached (see page 173).

Remove from the heat, skim and then allow the marmalade to cool for at least 15 minutes before potting and covering as usual (see page 173).

Mincemeats

The original mincemeat was, as its name suggests, a mixture of minced and shredded meat – beef, mutton or ox tongue – dried fruits, candied peel and spices. Since Elizabethan times it has been made into small 'minced' or 'shred' pies for Christmas and other festive occasions, as dried fruits and spices were expensive and therefore a special treat. The original Christmas mince pie was oval-shaped to represent the manger and contained a pastry baby, but this was outlawed, along with plum pudding, under the austerities of Oliver Cromwell's Commonwealth, for being far too rich and indulgent, as well as hinting of paganism.

The mince pie returned with the restoration of the monarchy, continuing with its original content until the eighteenth century, when it was discovered that a mixture of suet, spices and dried fruits could be combined with brandy or sack (sherry) some months before Christmas and stored perfectly satisfactorily in stone jars until needed. From this time, the mince pie lost its meat, retaining suet as its sole ingredient of animal origin.

Mincemeats can be made from any combination of dried and glacé fruits, nuts and spices, so it is worth experimenting. To keep well, the mincemeat should not have too high a proportion of apple or fresh ingredients to dried or candied fruits and sugar. The alcohol content and acid in the fruit juice also act as preservatives. Make sure the jars you store the mincemeat in are sterlised, as any impurities can cause fermentation. This can also be caused by incorrect storage, so find a cool place, such as an unheated room or a box in the garage. Store in the fridge if you have nowhere else suitable. Opened jars and part-used jars of mincemeat should be placed in the fridge and used up quickly.

A traditional method of preventing fermentation is to place the mincemeat, minus the brandy, in a cool oven (60°C/120°F) for 3 hours. Allow it to get completely cold before stirring in the brandy, potting and storing.

Traditional Mincemeat

A seventeenth-century recipe in the Dryden family papers at Canons Ashby, a National Trust property near Northampton, included minced ox tongue, beef suet and hard-boiled eggs with the dried fruit, apples, candied peel and spices. Rose- or orange-flower water and a little sack (sherry) or brandy was used to moisten the mincemeat. An eighteenth-century recipe from Erddig near Wrexham replaces the tongue with meat and omits the eggs and flower water. Mrs Scott, who gave the recipe to the Yorke family at Erddig, recommended using Seville oranges instead of sweet as they give an excellent flavour.

450g (1lb) eating apples
100g (3½oz) unblanched almonds
450g (1lb) seedless raisins
175g (6oz) sultanas
175g (6oz) currants
50g (1¾oz) mixed candied citron and lemon peel
50g (1¾oz) candied orange peel
100g (3½oz) suet
450g (1lb) soft brown sugar
Grated rind and juice of 2 lemons
Grated rind and juice of 2 sweet or Seville oranges
2 teaspoons ground mixed spice
2 teaspoons ground cinnamon
½ teaspoon freshly grated nutmeg
150ml (¼ pint) brandy

Peel, core and finely chop or grate the apples. Place the almonds in a basin and cover with boiling water. Leave for 5 minutes, then drain and slip off the skins. Place immediately in cold water, leave for a few minutes, then chop finely. Place all the ingredients, except the brandy, in a large bowl and stir well. Cover with a clean cloth and leave for two days in a cool place to allow the flavours to develop. Stir again very thoroughly, then stir in the brandy. Pack the mincemeat into sterilised jars and cover with clean lids or cellophane jam pot covers. Label and store for at least one month before using, to allow the flavours to mature. Use within three months.

Apricot and Hazelnut Mincemeat

If you dislike candied peel, this is the recipe for you. Dried figs make a delicious alternative to the dates if you want to experiment.

450g (1lb) cooking apples
225g (8oz) dates, stoned
450g (1lb) ready-to-eat dried apricots
100g (3½oz) natural glacé cherries, washed
50g (1¾oz) candied angelica
100g (3½oz) hazelnuts
50g (1¾oz) crystallised stem ginger
225g (8oz) suet
450g (1lb) sultanas
225g (8oz) seedless raisins
350g (12oz) soft dark brown sugar
Grated rind and juice of 1 lemon
Grated rind and juice of 1 orange
1 teaspoon ground mixed spice
1 teaspoon ground cinnamon
1 teaspoon ground mace
150ml (¼ pint) brandy

Peel, core and finely chop or grate the apples. Finely chop the dates, apricots, glacé cherries, angelica, hazelnuts and crystallised ginger. Mix all the ingredients together in a large bowl, cover with a clean cloth and leave to stand overnight. Stir again, then pot in sterilised jars. Cover and store for at least three months before using to allow the flavours to mature.

Cherry and Walnut Mincemeat

225g (8oz) natural glacé cherries, washed
100g (3½oz) whole candied peel
100g (3½oz) walnut pieces
700g (1½lb) cooked Bramley apple purée
1kg (2¼lb) mixed dried fruit
½ teaspoon freshly grated nutmeg
1 teaspoon ground cinnamon
Grated rind and juice of 1 lemon
Grated rind and juice of 1 orange
1 tablespoon ground mixed spice
225g (8oz) muscovado sugar
4 tablespoons brandy, sweet sherry or port

Coarsely chop the glacé cherries and peel. Finely chop the walnuts. Mix all the ingredients together in a large bowl. Cover with a clean cloth and leave overnight. Pack the mixture into sterilised jars, cover and store in a cool, dark place for up to six weeks before using.

Mincemeat Cooked in Cider

This recipe contains no suet and yet is full of fruity richness. A larger proportion of apples is used, as well as glacé cherries, and the mincemeat is cooked. Cider as well as rum is included, but you can omit this as you like. This mincemeat keeps extremely well.

400ml (¾ pint) medium cider
450g (1lb) soft dark brown sugar
1·8kg (4lb) cooking apples, peeled, cored and roughly chopped
450g (1lb) currants
450g (1lb) seedless raisins
100g (3½oz) natural glacé cherries, washed and chopped
100g (3½oz) blanched almonds, chopped
Grated rind and juice of 1 lemon
1 teaspoon ground mixed spice
1 teaspoon ground cinnamon
½ teaspoon ground cloves
2 tablespoons rum

Place the cider and sugar in a large pan and heat gently, stirring occasionally, until the sugar has dissolved. Add the apples to the pan, and stir in all the remaining ingredients except the rum. Bring the mixture slowly to the boil, stirring all the time. Reduce the heat, half-cover with a lid and simmer gently for about 30 minutes or until the mixture has become a soft pulp, stirring occasionally. Test for sweetness, adding more sugar if necessary. Remove from the heat and set aside until completely cold. Stir in the rum, then pack into sterilised jars, cover and store.

Pear and Fig Mincemeat

This recipe uses pears instead of apples as its base. You can substitute dates for the figs if you prefer, and walnuts for the almonds. Sherry, rum, whisky or cider can be used instead of brandy.

900g (2lb) cooking pears
450g (1lb) seedless raisins
450g (1lb) sultanas
225g (8oz) currants
225g (8oz) dried figs
100g (3½oz) whole candied orange peel
100g (3½oz) mixed candied citron and
   lemon peel
225g (8oz) blanched almonds
450g (1lb) demerara sugar
450g (1lb) suet
Grated rind and juice of 2 lemons
2 teaspoons ground mixed spice
1 teaspoon ground ginger
1 teaspoon ground cinnamon
½ teaspoon freshly grated nutmeg
150ml (¼ pint) brandy

Peel, core and finely chop or grate the pears. Chop the other fruit, peel and nuts. Place all the ingredients in a large china bowl. Cover with a clean cloth and leave to stand for two or three days. Stir the mincemeat thoroughly, then pack into sterilised jars, cover and store.

Freezer Mincemeat

700g (1½lb) cooking apples, peeled, cored and finely chopped
225g (8oz) currants
225g (8oz) sultanas
225g (8oz) seedless raisins
100g (3½oz) mixed candied citron and lemon peel
100g (3½oz) whole candied orange peel
75g (2¾oz) blanched almonds
450g (1lb) soft dark brown sugar
100g (3½oz) suet
1 teaspoon ground cinnamon
1 teaspoon freshly grated nutmeg
Grated rind and juice of 1 lemon
Grated rind of 1 orange

Blanch the apples in boiling water for 30 seconds, then drain very thoroughly in a colander and cool. Place the cooled apples in a large bowl. Chop the dried fruit, peel and nuts and add to the apples with all the remaining ingredients except the brandy or sherry. Stir thoroughly, then ladle the mixture into plastic containers and freeze for up to four months. Thaw overnight in the fridge, then add brandy or sherry before using.

Fruits Preserved in Alcohol

Bottling fruits in alcohol is one of the oldest, simplest and most delicious ways of preserving them. The luxury fruits grown in the eighteenth century, such as nectarines, peaches, apricots, cherries and grapes, were originally known as 'brandy fruits' because they were often preserved in brandy and a little sugar syrup ready for the dessert course, when they were usually served in sweetmeat glasses.

These preserves make the most luxurious desserts for a special occasion – traditionally at Christmas – and very acceptable presents, especially when packed in attractive airtight jars with decorative labels.

To Make Fruits in Alcohol

1. Fruits must be in perfect condition and just ripe; if over-ripe they will not keep their shape in the alcohol.
2. Do not be tempted to use cheap alcohol, as the alcohol content must be as high as possible to preserve successfully.
3. Granulated sugar is the usual choice of sweetener; brown sugar can be used for flavour and colour if preferred.
4. Pick over the fruit, discarding leaves, stalks, stones, pips, peel and cores as necessary, then wash and dry.
5. Soft fruit, such as berries, blackcurrants, apricots and peaches can be used raw, while firmer ones such as apples and hard plums, may require light cooking first.
6. Pack the fruits into sterilised jars (see page 7), either in layers with sugar or in sugar syrup. Warm the jars if the ingredients used are hot.
7. Pour the chosen alcohol over the fruits to cover them completely, making sure that there are no air pockets between them. (The amount of alcohol needed may vary, depending on the size of the fruits and the amount of sugar used.)
8. Spices such as cloves, cinnamon sticks and allspice berries can be added to the fruits for extra flavour.
9. Seal the jars tightly and leave the fruits in a cool, dark place for at least one month before using, to allow the flavours to develop. Shake the jars from time to time to blend and dissolve the sugar and flavourings.

As long as the fruits are covered in the alcohol, they should keep for about one year, and the flavour will improve over this time.

Apricots in Amaretto

As well as supporting fan-trained fruit, the walls of old kitchen gardens were built as 'hot walls', with flues at intervals to ripen apricots and peaches and to protect the blossom from late frosts. Such walls can be seen in various National Trust properties such as the delightful walled garden at Greys Court, near Henley-on-Thames in Oxfordshire, or Westbury Court in Gloucestershire. Peaches or nectarines may be used instead of apricots in this recipe.

450g (1lb) fresh apricots
100g (3½oz) granulated sugar
Amaretto to cover, about 300ml (½ pint)

Wipe the apricots if necessary and discard any leaves and stalks. Leave the fruit whole. Prick them all over with a fine sterilised needle or a cocktail stick, so that the Amaretto can penetrate the flesh, and then pack into sterilised jars with the sugar.

Pour over enough Amaretto to cover the apricots by 1cm (½in), making sure there are no pockets between the fruits. Seal tightly and then shake well so that the sugar can start to blend and dissolve in the alcohol.

Keep in a cool, dark place for at least two months before eating. Shake from time to time during the first week of storage to make sure all the sugar dissolves in the alcohol.

VARIATION
Brandied Apricots
Use brandy instead of Amaretto.

Dried Apricots in Eau-de-Vie

This recipe works equally well with fresh or dried cherries, dried pears or with Muscatel raisins.

450g (1lb) good-quality ready-to-eat dried apricots
300ml (½ pint) cold water
225g (8oz) granulated sugar
3 sprigs of fresh lemon verbena
About 300ml (½ pint) eau-de-vie

Remove any stalks and leaves from the apricots, then place in a 1-litre (2-pint) sterilised preserving jar – the fruit should fill just over half the jar. Place the water and sugar in a pan and stir over gentle heat until the sugar has dissolved. Bring to the boil, then simmer for 5 minutes. Add the lemon verbena sprigs to the apricots, then pour in the hot sugar syrup. Pour in enough eau-de-vie to come to the rim of the jar, making sure that the fruit is covered. Seal the jar tightly and label, then leave in a cool, dark place for four to eight weeks before using.

Damsons in Mulled Wine

These make a delicious dessert, especially at Christmas.

75cl bottle red wine
4 whole cloves
6 whole cardamom pods, crushed
2 cinnamon sticks
225g (8oz) granulated sugar
900g (2lb) damsons

Pour the red wine into a pan and add the spices and sugar. Heat gently, stirring until the sugar has completely dissolved, and then bring to the boil. Reduce the heat and simmer for 20 minutes.

Wash and dry the damsons, discarding any leaves and stalks. Prick all over with a fine sterilised needle or cocktail stick and then add to the wine mixture.

Simmer very gently for about 10 minutes or until the damsons are tender. Remove the pan from the heat and allow to cool. Using a slotted spoon, pack the damsons into sterilised jars, removing as many stones as you can. Strain the wine syrup and pour it over the damsons to cover them by 2.5cm (1in). Seal the jars and label. Keep in a cool, dark place for one month before using to allow the flavours to develop, but use within three months.

Spiced Fruit and Walnut Compote in Calvados

6 firm ripe pears, peeled and cored
225g (8oz) caster sugar
4 whole cloves
1 bay leaf
1 cinnamon stick
1 vanilla pod, split
1 star anise
5 whole black peppercorns
½ teaspoon coriander seeds
450g (1lb) ready-to-eat prunes
225g (8oz) mixed ready-to-eat
    dried fruit, such as apricots, apples
    and figs
100g (3½oz) walnut halves (optional)
Calvados or apple/cider brandy

Quarter or halve the pears, depending on size. Put them in a pan with the sugar and spices and enough water to just cover the fruit. Cover the pan with a lid and heat gently, stirring, until all the sugar has dissolved. Simmer very gently for 10–15 minutes, or until the pears are tender. Remove from the heat, add the prunes, dried fruits and nuts, then leave to cool. Using a slotted spoon remove the fruit and nuts from the syrup and place in sterilised jars. Boil the reserved syrup over a high heat until it has thickened and reduced to 300ml (½ pint). Pour over the fruit, and then pour in enough Calvados to come to within 2·5cm (1in) of the tops of the jars. Seal and label, then keep in a cool, dark place for at least one month before using.

Black Cherries in Port

The aristocrats of the traditional English fruit garden were the wall-trained fruit – the peaches, apricots and cherries. Try using Kirsch, brandy, rum or Amaretto for a change in this recipe. After they have matured for a month, the cherries can be decanted into pretty jars if you are giving them as presents.

700g (1½lb) fresh black cherries
225g (8oz) granulated sugar
Good-quality port to cover

Wash and dry the cherries, discarding any leaves and stalks. Stone with a cherry-stoner (you can buy these cheaply and they work brilliantly). Prick the fruit all over with a fine sterilised needle or a cocktail stick, which allows the port to permeate the skins and flesh, preventing the cherries from shrivelling. Layer the cherries and sugar in sterilised preserving jars, filling them to 2.5cm (1in) below the rim. Pour in port to cover, then seal the jars. Store in a cool, dark place for at least one month, shaking gently every day during the first week.

Brandied Peaches

These make a lovely dessert to enjoy at Christmas, or any other time. They also make a super present. If you preserve the peaches in August when they are at their best (and cheapest) you will get the best result.

6 ripe, juicy peaches
Juice of 1 lemon
6 whole cloves
400ml (14fl oz) cold water
350g (12oz) granulated sugar
5cm (2in) piece of cinnamon stick
¼ teaspoon ground mace
About 150ml (¼ pint) brandy

Cover the peaches with boiling water for 1–2 minutes and then plunge into iced water. Remove from the water and peel off the skins. Brush the peaches all over with lemon juice to stop them going brown and stick a clove in each one.

Put the water, sugar and spices into a pan large enough to take the peaches. Bring to the boil slowly, stirring frequently until the sugar has dissolved. Add the peaches and cook for 5–10 minutes, or until tender. Remove the peaches with a slotted spoon and pack into hot, sterilised, wide-necked jars. Stir the brandy into the cooking liquor and pour over the fruit to cover completely.

Seal and store in a cool, dark place for at least a week before eating.

Peach Compote with Brandy

I was inspired to make this recipe after my husband brought back a delicious jar from France, which was particularly good with meringues and our local Cornish clotted cream! The compote would be excellent in pies, cobblers and crumbles too.

500g (1lb 2oz) ripe, juicy peaches
225g (8oz) granulated sugar
Juice of ½ lemon
1 tablespoon brandy

Cover the peaches with boiling water for 1–2 minutes and then plunge into iced water. Remove from the water and peel off the skins. Halve the peaches and remove and discard the stones and then roughly chop the flesh.

Place the peach flesh in a pan with the sugar, lemon juice and brandy and bring slowly to the boil, stirring until all the sugar has dissolved. Simmer for 5–10 minutes, or until the mixture has reduced by about 10 per cent.

Pack into hot, sterilised jars and seal. Keep in the fridge and use as you wish. The compote will keep for at least six months.

Squiffy Prunes

Buy giant-sized, ready-to-eat prunes for this recipe. It makes a lovely dessert served with cream or ice cream, or a good appetiser wrapped in a thin rasher of bacon and grilled. The liquor can be drunk as a dessert wine.

Giant ready-to-eat prunes
Tawny port to cover

Pack the prunes into sterilised jars and cover completely with port. Seal and store in a cool, dark place for at least two months before using.

Oranges in Brandy

The brandy need not be of the best quality, but do not be tempted to use very cheap brandy either. Try satsumas, tangerines or mandarins instead of oranges. Kumquats can also be used, but should be left whole and pierced all over with a sterilised needle to allow the brandy to penetrate the flesh. Experiment with different liqueurs and spirits, such as Cointreau, Kirsch or rum, if you wish.

12 small oranges
350g (12oz) granulated sugar
300ml (½ pint) water
2 teaspoons allspice berries
About 225ml (8fl oz) brandy

Using a potato peeler, peel the rind very thinly from 3 of the oranges, with removing any pith. Cut the rind into fine strips and blanch in boiling water for 30 seconds, then drain and leave to cool. Using a sharp knife, cut off the peel and pith from the other 9 oranges and discard. Put 175g (6oz) sugar and the water in a wide, shallow pan and heat gently, stirring until all the sugar has dissolved. Add the oranges and allspice berries and poach for 5 minutes, turning once.

Spoon the oranges into a colander placed over a smaller pan and leave to drain for 10 minutes, then put on one side to get completely cold. Pour the rest of the poaching syrup into the small pan and add the remaining sugar. Stir over a low heat until the sugar has completely dissolved, then boil rapidly until the temperature rises to 110°C (225° F); use a sugar thermometer. Remove the pan from the heat and pour the syrup into a measuring jug. Leave to get completely cold, then stir in an equal amount of brandy.

Pack the oranges into sterilised preserving jars, filling them almost to the top. Add the reserved orange rind to the syrup, then pour over the fruit to cover completely. Seal with airtight, acid-proof screw caps. Label and store in a cool, dark place for at least two months before using.

Fruit and Flower Drinks

For centuries, fruit, herb and flower-flavoured drinks have been made for both pleasure and for reasons of health. King Henry III is said to have enjoyed wine flavoured with scented flowers, while his son, Edward I, had a taste for sage wine.

In the seventeenth century, ale flavoured with elderberries was drunk in preference to port, and raspberry juice was added to wine to make the popular raspberry sack. Citrus fruit juice, known for its protection against scurvy, was mixed with sack (sherry) or brandy to make the first lemonade and orangeade.

The eighteenth- and nineteenth-century 'compleat housewife' had to be able to produce numerous waters, ratafias and cordials from spirits flavoured with red roses, poppy petals and the leaves and flowers of rosemary, marjoram and lavender; with fruits such as cherries and raspberries; and with nuts such as almonds and apricot kernels.

All the recipes I have chosen are simple to make; producing your own sloe gin or apricot brandy will give you enormous pleasure as well as making you into a twenty-first-century 'complete housewife'.

When buying the required alcohol, choose the highest proof or percentage of alcohol you can find as nothing can grow in pure alcohol, so the strongest brand available will give the best results.

Apricot Brandy

Peaches and nectarines can be used instead of apricots, but always use ripe fruit. Drink as a liqueur or use to flavour apricot puddings or cakes.

450g (1lb) fresh ripe apricots
225g (8oz) granulated sugar
About 600ml (1 pint) brandy

Wash the apricots, discarding any leaves and stalks, then dry them. Cut into small pieces, removing and reserving the stones. Crack the stones with a nutcracker, and remove the kernels. Crush the kernels.

Put the chopped apricots into a large sterilised preserving jar and add the crushed kernels and the sugar. Pour in the brandy and seal tightly. Store in a cool, dry place for one month, shaking frequently.

Strain through a muslin-lined sieve overnight, and then pour the apricot-flavoured brandy into sterilised bottles with a screw cap. Seal tightly and store for another three months before using.

The leftover apricot pieces can be eaten as a special dessert on their own, mixed into a fruit salad or used as a filling for sweet pancakes or meringues.

Erddig Lemon Brandy

This recipe, from the Erddig archives, was provided by a Mrs Harvey. Drink this as a liqueur or use as a flavouring for custards, puddings, cakes or sauces. Lemon Brandy was the 'secret' ingredient in the original Bakewell pudding and was a common flavouring in the eighteenth and nineteenth centuries. Whisky can be used instead of brandy.

3 large lemons
600ml (1 pint) brandy
100g (3½oz) caster sugar

Scrub and dry the lemons. Using a potato peeler remove the peel without removing any pith. Put the peel in a clean jar. Squeeze the juice from the fruit and add it to the jar with the brandy and sugar. Seal the jar and shake well to help dissolve the sugar. Store in a cool, dark place for one to two weeks, shaking the jar occasionally.

Strain the brandy through muslin or fine cotton and pour it into sterilised bottles. It is now ready to drink.

VARIATION
Orange Brandy
Use oranges instead of lemons.

Sloe Gin

Sloes are the fruit of the spiky blackthorn, one of the wild ancestors of our many varieties of cultivated plum, common in hedgerows, moorland and open woodland. They are ripe in the autumn, but you will have to beat the numerous birds that love to eat them. Tradition has it that the best time to pick sloes is after the first frost of autumn has swollen and softened them slightly, but with climate change you could be waiting for ever. I have found that picking them at their best and freezing them overnight has the same effect and they then yield more juice.

There are numerous recipes for this traditional country drink, which is usually opened with great ceremony at Christmas time. The amount of sugar varies considerably from recipe to recipe and it really is a matter of individual taste – you will have to experiment! I recommend you start with 100g (3½oz).

Sloes impart a wonderful, warming fruitiness and colour to the gin, which makes a delicious addition to numerous puddings. Try adding to poached plums or quinces, or to fruit pies or crumbles.

450g (1lb) sloes
50–175g (1¼–6oz) granulated sugar
About 700ml (1¼ pints) gin

Wash the sloes, discarding any leaves and stalks. Freeze overnight.

Next day, thaw the sloes enough to prick the flesh several times with a sterilised needle, then drop them into a 1·5-litre (2¾-pint) preserving jar. Add the sugar, then pour in the gin, reserving the bottle for later use. Seal tightly and invert several times to distribute the sugar and start it dissolving. Leave to infuse in a cool, dark place for at least three months, shaking every day for a month.

After three months, strain the gin through a muslin-lined sieve and decant back into the original bottle. Screw on the original cap and drink straight away or, better still, leave to mature for next Christmas. The sloes can also be eaten.

Variations
Sloe Gin Cocktails
Sloe gin is usually drunk neat, but can be turned into a cocktail. Half fill a glass with crushed ice, add a wine glass of sloe gin and a dash of orange bitters.
To make Sloe Gin Fizz shake one measure of sloe gin with a teaspoon of lemon juice, then add soda water to taste. A generous glug of sloe gin added to a glass of Champagne or sparking white wine makes the most wonderful aperitif.
Drop two or three of the gin-soaked sloes into the glass to make it look even more elegant.

Fruit and Flower Drinks

Damson Gin

Damsons can often be found growing wild in hedgerows. They can be used to make a warming winter tipple. Drink as a liqueur or add to poached fruit, fruit pies, tarts and crumbles or to add zing to fruit sauces.

450g (1lb) damsons
100g (3½oz) granulated sugar
About 600ml (1 pint) gin

Wash the damsons well, discarding any damaged fruit as well as leaves and stalks. Dry the fruit, then prick several times with a sterlised needle. Drop into a large preserving jar and add the sugar. Pour in the gin, reserving the gin bottle for later use, then seal tightly and give everything a good shake to start the sugar dissolving. Leave to infuse in a cool dark place for at least three months, shaking every day for a month.

After three months, strain the gin through a muslin-lined sieve and decant back into the original bottle. Screw on the original cap; it is now ready to drink.

VARIATION
Damson or Sloe Vodka
Use vodka instead of gin and proceed as above.

Imperial Pop

This excellent recipe for ginger beer dates back to the early nineteenth century. A number of the National Trust restaurants make their own using a similar recipe.

1 lemon
450g (1lb) caster sugar
25g (1oz) cream of tartar
25g (1oz) fresh root ginger
5 litres (9 pints) boiling water
25g (1oz) fresh yeast or 1 sachet of
    dried yeast

Use a potato peeler to pare the rind from the lemon, avoiding the pith. Squeeze out the juice. Put into a large china bowl with the sugar and cream of tartar. Using a steak hammer or rolling pin, severely bruise the ginger without peeling it and add to the bowl. Pour over the boiling water and stir well. When the water is just tepid, stir in the yeast. Cover and leave overnight to ferment.

Next day, skim and pour through a muslin-lined nylon sieve into screw-capped plastic fizzy drink bottles. Leave for at least two to three days before drinking, checking after 36 hours to make sure that the ginger beer is not too fizzy. If it is in danger of exploding out of the bottles, loosen the caps very slightly. Serve well chilled.

Victorian Cherry and Raspberry Ratafia

In the eighteenth century ratafia was the name for a liqueur made from peach or apricot kernels and brandy. A recipe of 1765 in the Yorke family papers at Erddig, a National Trust property near Wrexham, calls for 'a Quart of the Best Brandy' and 'a Hundred Apricot Kernalls'.

By Victorian times ratafia was often used to describe any drink that was based on brandy, such as this recipe, which is flavoured with cherries, raspberries and spices. Drink as a liqueur, or use as a flavouring for custards, creams, cakes and sauces, or pour over fresh berries.

700g (1½lb) ripe black cherries
225g (8oz) fresh raspberries
225g (8oz) granulated sugar
About 600ml (1 pint) brandy
Large sprig of fresh coriander
1 cinnamon stick

Remove sufficient stones from the cherries to give 40g (1½oz) and reserve them in a covered container in the fridge. Put all the cherries into a bowl and gently mash with a potato masher, then tip into a large sterilised jar with the raspberries. Cover with the lid and leave for three days, stirring two or three times daily.

On the fourth day, crack open the reserved cherry stones with a nutcracker and remove the kernels. Blanch them by pouring over boiling water; leave them to stand for 1 minute, then drain and skin them. Add the sugar to the fruit and stir until completely dissolved. Stir in the kernels with the brandy, coriander and cinnamon. Cover again and leave in a cool, dark, airy cupboard to infuse for one month.

Pour the brandy through a muslin-lined nylon sieve into a bowl, discarding the coriander and cinnamon. Leave overnight until all the ratafia has dripped through the sieve.

Next day, pour the ratafia into sterilised bottles. Seal tightly, and store in a cool, dry place for another three months before drinking.

Fruit and Flower Drinks

Fresh Lemonade

Lemonade was a French invention, but English drinkers liked to add an equal quantity of white wine to make it rather more heady! Many of the National Trust restaurants make their own versions of this wonderfully refreshing drink. This is a quick and easy recipe, which is less sweet than most shop-bought lemonades and also free from artificial colours and flavours.

3 large lemons
About 850ml (1½ pints) boiling water
50–75g (1¾–2¾oz) caster sugar

Scrub the lemons, then slice roughly and purée them coarsely in a food processor or blender, adding about 450ml (¾ pint) of the water and about a third of the sugar. Strain through a nylon sieve, then repeat the process twice more using the pulp remaining in the sieve. Add extra water and sugar to taste. Chill well, then serve decorated with mint, borage flowers and lemon slices.

To serve, pour a little cordial into a chilled glass and fill up with cold water, tonic or sparkling mineral water. Add ice and a slice of lemon.

Elderberry Liqueur

Pick the elderberries in late September when they are a deep, almost black, purple to make this rich British liqueur.

900g (2lb) elderberries
600ml (1 pint) brandy
350g (12oz) granulated sugar
1 teaspoon ground allspice
½ cinnamon stick
1 whole clove
Large pinch of ground mace

Wash the elderberries and discard any leaves. Using a fork, strip them off the stalks into a bowl. Lightly mash the fruit and stir in the other ingredients. Ladle the mixture into a large sterilised preserving jar and seal tightly. Leave in a cool, dark place for one month.

Strain the liqueur through a muslin-lined nylon sieve, discarding the elderberries and spices. Pour it into sterilised bottles with screw caps. Seal and store in a cool place for a further three months before drinking.

Hawthorn Flower Liqueur

Hawthorn flowers, also known as may, appear in May or early summer, hence the name. They should be gathered when freshly opened, early in the morning of a dry sunny day. Only the petals are used to make this liqueur, which has a delicate almond flavour. It makes a pleasing drink or it can be used to flavour puddings, fruit compotes, jellies, cakes and biscuits.

100g (3½oz) hawthorn flower petals
About 600ml (1 pint) brandy
175g (6oz) caster sugar
4 tablespoons cold water

Make sure that the flower petals are insect-free, and then place in a large sterilised preserving jar. Cover with the brandy, seal and store in a cool, dark place for two to three months.

When you are ready to finish the liqueur, put the sugar and water into a pan. Heat gently, stirring frequently, until the sugar has completely dissolved and then bring to the boil. Boil for about 1 minute and then leave until completely cold.

Strain the brandy through a muslin-lined nylon sieve into a large jug, then add the sugar syrup. Stir well and then pour into sterilised bottles with screw caps. Seal tightly and store in a cool, dark place, until required.

Elderflower Champagne

Gather the elderflower blossom on a dry sunny day and choose newly opened flowers to guarantee the maximum flavour. Use as quickly as possible, to retain freshness.

2 lemons
700g (1½lb) granulated sugar
2 tablespoons white wine vinegar
About 24 fresh elderflower heads
5 litres (9 pints) cold water

Put the juice and rind of 1 lemon into a very clean bowl or bucket that will hold 5 litres (9 pints). Slice the second lemon and add to the bowl with the sugar and vinegar. Make sure the blossom is insect-free and then snip the individual flowers from their stems and add to the bowl. Bruise well with a potato masher and then pour in the water. Stir well and then cover with a clean cloth and leave to stand in a cool place for two to four days, stirring occasionally.

Strain twice through scalded muslin and then bottle. Leave to stand in a cool place for 10–14 days before drinking, very cold, in well-chilled glasses. Once the champagne is ready, drink within a month as its flavour will deteriorate.

Elderflower Cordial

About 15 large fresh elderflower heads
1 large lemon
25g (1oz) citric acid
1kg (2¼lb) granulated sugar
700ml (1¼ pints) boiling water

Make sure the elderflowers are insect-free and then snip the main stalk from each head and let the flowers drop into a large mixing bowl. Wash the lemon, slice it and then add to the bowl with the citric acid and sugar. Pour the boiling water over the mixture and stir until the sugar has dissolved. Cover with a clean cloth and leave for five days, stirring daily.

When the cordial is ready to be bottled, line a colander with two layers of scalded muslin and set over another mixing bowl. Tip the elderflower mixture into the colander and let the cordial strain through. Discard everything in the muslin and then pour the cordial into small, sterilised bottles with screw caps and store in the fridge for up to three months.

To serve, pour a little cordial into a chilled glass and fill up with cold water, tonic or sparkling mineral water. Add ice and a slice of lemon.

Lemon Barley Water

Barley water was originally drunk as a medicine. A recipe dating to 1685 from Erddig, a National Trust property near Wrexham, is described as 'A Dainty Cooling Drink for a Hot Fever'. Later it became the drink of genteel Victorian ladies and is forever linked with tennis and croquet parties.

100g (3½oz) pearl barley
50g (1¾oz) white sugar cubes
4 large lemons
1·2 litres (2 pints) boiling water

Rinse the pearl barley in a sieve under cold running water and drain well. Place in a pan and pour in cold water to cover, then bring to the boil. Reduce the heat and simmer gently for 5 minutes. Pour into a sieve and rinse again, then put into a large jug or bowl. Rub each sugar cube over the rinds of the lemons to extract their oils, then add the cubes to the barley. Pour in the boiling water and stir until the sugar has dissolved. Cover with a clean cloth and leave to infuse for 3 hours or until cold.

Squeeze the juice from the lemons, add it to the barley water and strain through a nylon sieve. Cover and chill for at least 1 hour before serving with plenty of ice and lemon slices.

Variations
Lime Barley Water
Substitute 6–8 limes (depending on size) for the lemons.

Orange Barley Water
Substitute 4 oranges (depending on size) for the lemons and use 25g (1oz) sugar cubes.

St Clement's Cordial

This lovely recipe was given to me by a Cornish friend who frequently wins prizes in local shows.

4 large juicy oranges
2 lemons
900g (2lb) granulated sugar
1·2 litres (2 pints) boiling water
25g (1oz) citric acid
15g (½oz) tartaric acid

Mince or process the fruit coarsely and place in a large bowl. Add the sugar and the boiling water and stir well. Cover and leave to steep for 24 hours.

Stir in the citric and tartaric acids. Strain at least three times through a nylon sieve, then pour into sterilised bottles. Cover tightly and store in the fridge. Dilute to taste and serve chilled, with orange slices, lemon balm and borage flowers.

Raspberry Shrub

This excellent summer drink tastes equally good served hot in winter. Dilute it with boiling water and sip it, as the Victorians did, to soothe sore throats and to ward off colds. It also makes a good sauce for sponge puddings, ice creams or rice puddings, or it can be used in sweet and sour dishes.

350g (12oz) granulated sugar
600ml (1 pint) raspberry vinegar

Heat the sugar and vinegar together gently until the sugar has completely dissolved. Bring to the boil, then boil for 10 minutes. Pour into sterilised bottles, cover and leave to cool. Keep in the fridge for up to six weeks.

Dilute to taste with still or sparkling water, or soda water. For a more sophisticated drink, top up 1 teaspoon of raspberry shrub with sparkling white wine. To make milkshakes, whisk 1 tablespoon into a glass of chilled milk.

VARIATIONS
Strawberry, Redcurrant, Loganberry or Blackberry Shrub
Substitute the appropriate fruit vinegar for raspberry.

Index

Apple Butter, Spiced 151
Apple Chutney (Cliveden's) 12
Apple Curd (Mayfield) 166
Apple Geranium Jelly 131
Apple Jelly 130
Apple Lemon Jelly, 131
Apple, Onion and Sage Chutney 13
Apricot and Ginger Chutney 24
Apricot and Hazel Mincemeat 188
Apricot Brandy 206
Apricot Jam 119
Apricot Pastilles 163
Apricots (Dried), in Eau-de-Vie 196
Apricots in Amaretto 195
Autumn Chutney 22

Banana and Apple Chutney 21
Bay Salt, 102
Beetroot and Ginger Relish 26
Beets (Pickled Baby) with Horseradish 58
Black Cherries in Port 199
Black Cherry Conserve 111
Blackberries, Spiced 60
Blackberry and Apple Jam 116
Blackberry Curd 167
Blackberry, Apple, Orange and Juniper Butter 152
Blackberries, Spiced 61
Blackcurrant Jam 115
Blackcurrant Syrup 88
Blueberry Conserve 120
Borage Sugar, 101
Bramble and Apple Butter 150
Bramley (Fresh) and Date Relish 14

Cherry and Raspberry Ratafia (Victorian) 212
Cherry and Walnut Mincemeat 188
Chillies, Pickled, 57
Chutneys, Relishes and Sauces 8

Cinnamon Sugar 99
Coarse-cut Marmalade 181
Cordial (St Clement's) 222
Courgette Chutney 28
Crab Apple and Clove Jelly 132
Crab Apple Cheese 154–155
Crab Apple Pickle 62
Cranberry and Apple, Spiced Cheese 156
Cucumber Pickle 68
Cumberland Sauce 50
Curried Fruit Relish (Cragside's) 25

Damson (Sweet) Pickle 64
Damson Cheese 157
Damson Chutney 29
Damson Gin 210
Damson Jelly 138–139
Damsons in Mulled Wine 196

Eggs, Pickled 83
Elderberry Chutney 16
Elderberry Liqueur 216
Elderflower Champagne 218
Elderflower Cordial 220
Elderflower Syrup 105

Fig Relish 31
Flavoured Sugars, Salts and Syrups 86
Fruit and Walnut Compôte in Calvados (Spiced) 197
Fruit and Flower Drinks 204
Fruit Butters 150
Fruit Butters, Cheeses, Pastilles, Leathers and Curds 148
Fruit Cheeses 154
Fruit Chutney (Father's Special) 49
Fruit Curds 164
Fruit Leathers 160
Fruit Pastilles 161
Fruit Syrups 80
Fruits Preserved in Alcohol 192

Gooseberry Chutney 32
Gooseberry Curd 168

Granny's Hot 20
Grapefruit and Lemon Marmalade 174
Green Gooseberry Cheese 158
Greengage and Orange Jam 126

Hawthorn Flower Liqueur 217
Herb and Flower Syrups 104
Herb Salts 102
Herb Sugars 100
High Dumpsie Dearie Jam 124
Horseradish Vinegar 58

Indian Pickle 65

Jams and Jellies 106
Jellies 128

Kea Plum Jam 127

Lavender Sugar 100
Lemon Barley Water 221
Lemon Brandy 207
Lemon Curd (Trelissick) 164
Lemon Marmalade 175
Lemonade, Fresh 214
Lemons, Preserved 69
Lime Marmalade 182

Mango Chutney 19
Marmalades 170
Marmalade (Dark Coarse-cut), 181
Marrow (Fresh) Relish 36
Marrow, Red Tomato and Date Chutney 35
Melanie's Bramble Jelly 136
Mincemeat (Freezer) 191
Mincemeat (Traditional) 186
Mincemeat Cooked in Cider 189
Mincemeats 184
Mint and Apple Jelly (Piquant), 134
Mint or Sage Jelly 147
Mulberry Jelly 138
Mulled Wine Syrup 91
Mushroom Ketchup 52
Mushrooms, Pickled 81

Onions (Sweet) Pickled 73
Orange and Rosemary Jelly 135
Orange or Lemon Sugar 95
Orange Slices, Spiced 76
Oranges in Brandy 202

Peach Compote with Brandy 201
Peaches, Brandied 200
Peaches, Sweet Pickled 75
Pear and Fig Mincemeat 190
Pear, Orange and Ginger Chutney 23
Pears, Ginger Pickled 79
Peppercorn and Rosemary Salt 103
Piccalilli (Cotehele's Sweet) 67
Pickles 54
Pineapple Sage Syrup 104
Plum (Spiced), Jelly 142
Plum Butter 153
Pop, Imperial 211
Prunes, Squiffy 201
Prunes, Spiced 78
Pumpkin and Raisin Chutney 37

Quince Cheese 159
Quince Comfits 161
Quince Jelly (Old-fashioned) 144
Quinces, Spiced 74

Raspberry Conserve (Uncooked) 123
Raspberry Jam 122
Raspberry Shrub 223
Red Cabbage, Pickled 63
Red Chilli Jam 121
Red Gooseberry and Elderflower Jam 125
Red Onion Marmalade 39
Red Pepper, Tomato and Lemongrass Chutney 42
Red Tomato and Ginger Chutney 48
Redcurrant Jelly (My Mother's) 140
Rhubarb (Wimpole's

Spring), Melon and Ginger Relish 45
Rhubarb and Coriander Chutney 44
Rhubarb and Ginger Jam 112
Rhubarb and Orange Butter 153
Rhubarb and Orange Jelly 143
Rhubarb and Rose Petal Jam 114
Rhubarb Sauce 47
Rosehip Syrup 92
Rosemary, Thyme, Basil or Marjoram Jelly 134
Rowan and Crab Apple Jelly 133
Runner Bean Pickle 70

Sage, Pineapple Syrup 104
Seville Orange and Apricot Chutney 40
Seville Orange Jelly Marmalade 180
Seville Orange Marmalade (Daddy's) 178
Sloe Gin 208–209
Spiced Blackberries 61
Spiced Blackberry Chutney 16
Spiced Fruit and Walnut Compote 197
Spiced Plum and Lime Chutney 30
Spiced Vinegar 61
Strawberry and Orange Curd 169
Strawberry and Raspberry Leather 160
Strawberry Jam 110

Tangerine Marmalade 176
Tarragon Salt 102
Three Fruit Marmalade 177
Tomato Ketchup 53

Vanilla Sugar 96
Vegetable, Sweet Pickle 82
Vinegar (Sweet Pickling) 62
Vinegar, Spiced 61
Walnuts (Green), Pickled 84